IRELAND'S
CIVIL ENGINEERING
HERITAGE

RONALD COX is a Research Associate in the Department of Civil, Structural & Environmental Engineering at Trinity College Dublin. Founder of the Centre for Civil Engineering Heritage, he is Chairman of the Engineers Ireland Heritage Society, and on the Board of the Irish Architectural Archive. Major publications include *Ireland's Bridges* (2002), and *Engineering Ireland* (2006).

PHILIP DONALD is a retired civil engineer whose career embraced the construction of major projects, including a dam in Argentina; and in Northern Ireland, the Foyle Bridge and the Belfast Cross-Harbour Bridges. He is a past chairman of ICE (NI). Publications include *Mud on my Boots: The Memoirs of a Civil Engineer* (2009).

Silent Valley Reservoir, County Down

IRELAND'S
CIVIL ENGINEERING
HERITAGE

EDITORS:
RONALD COX, PHILIP DONALD

CONTRIBUTORS:
RONALD COX, PHILIP DONALD, DERMOT O'DWYER, FRANK ROBINSON

The Collins Press

First published in 2013 by
The Collins Press
West Link Park
Doughcloyne
Wilton
Cork

A catalogue record for this book is available from the British Library

ISBN: 978-184889-1708

Design and typesetting by Burns Design
Typeset in Baskerville
Printed in Poland by Białostockie Zakłady Graficzne SA

COVER PHOTOGRAPHS

Front (clockwise from top left): water tower on UCD campus; Foyle Bridge, Derry (Esler Crawford Photography); Howth Harbour Lighthouse; Inistioge Bridge.

Back (from top): Craigmore Viaduct, Newry; southern approach to the Jack Lynch road tunnel, Cork (Cork City Council); 29th lock on the Grand Canal (Fred Hamond); train shed at Heuston Station, Dublin.

Background image: Rathmore rail tunnel, Cork.

CONTENTS

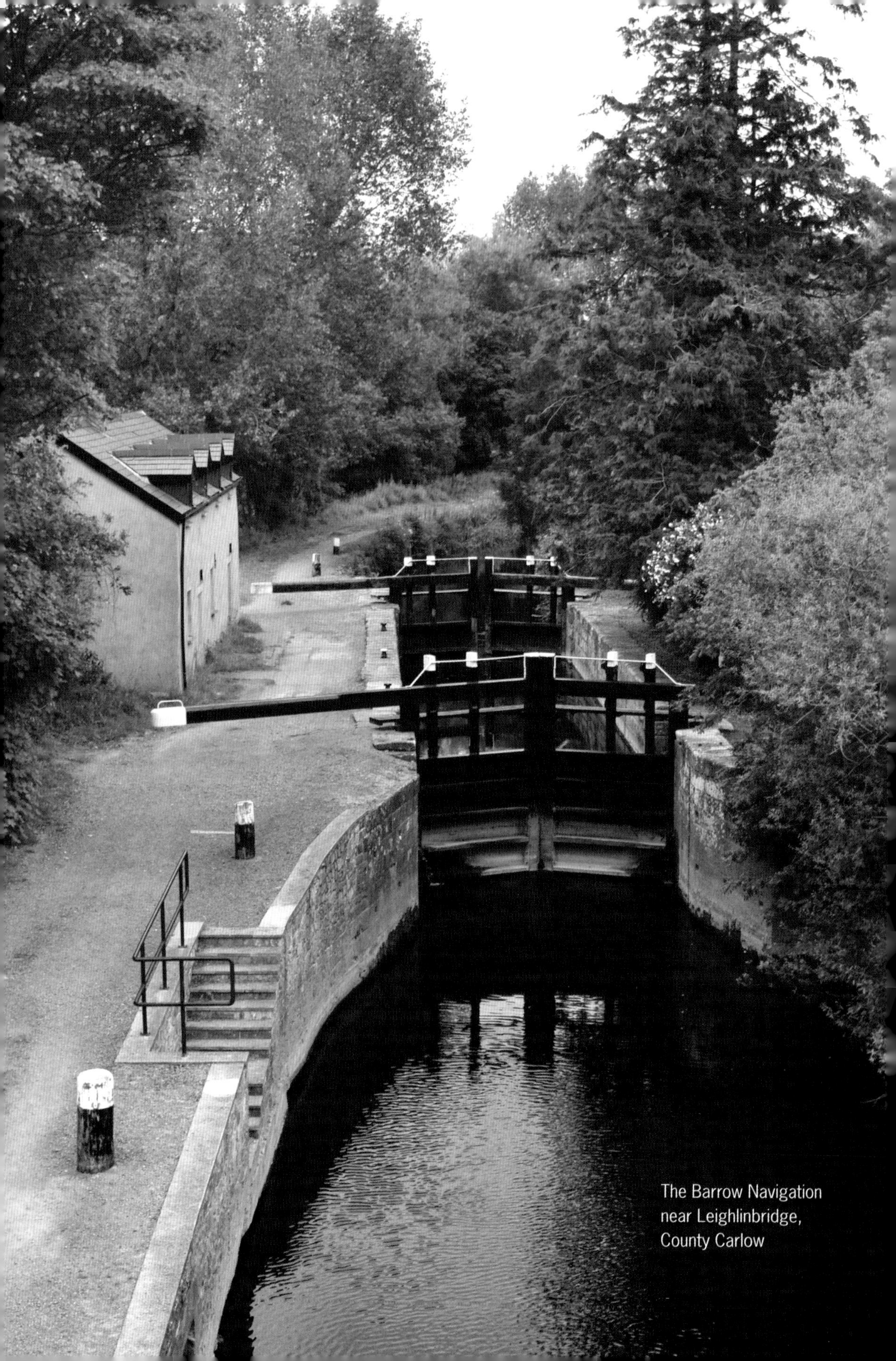

The Barrow Navigation
near Leighlinbridge,
County Carlow

PREFACE

Ireland has a significant heritage of civil engineering structures and works, and the skills of past engineers are in evidence throughout the island in the variety of bridges, canals, reservoirs, harbours, roads, railways and other societal infrastructure.

A number of civil engineering heritage books have been published, notably volumes by Thomas Telford Publishing, and more recently by Phillimore, covering various regions of Britain and the island of Ireland. These publications seek to make available to a wide readership information about civil engineering works, in the hope that they will encourage interest in and an understanding of the wide range of expertise that has contributed to our built heritage.

In recent years, there has been a growing awareness of the value of Ireland's built heritage, coupled with ongoing efforts to conserve the best examples for the appreciation of future generations. It is hoped that this book will be of assistance in advancing the case for such conservation.

Much of the information in these books, particularly in the Gazetteer section, has been extracted from the records of civil engineering works compiled by Ronald Cox of the Heritage Society of Engineers Ireland (EI) and Michael Gould of the Panel for Historical Engineering Works of the Institution of Civil Engineers (ICE). The records have, in a number of cases, been updated by current members of PHEW, Dermot O'Dwyer and Frank Robinson. The works covered by these records have been selected for their technical interest, innovation, association with leading engineers or contractors, rarity or visual attraction.

The records of civil engineering works in Ireland are currently held in the Civil Engineering Archive in Trinity College Dublin and the ICE Archives and may be consulted by the public for research purposes. The relevant Historical Engineering Work (HEW) registration number, where one exists, is provided for the entries in the Gazetteer to assist researchers in obtaining further information. A national grid map reference is also given for each site. To assist with a fuller appreciation of this heritage, readers are provided with an extensive glossary of technical terms.

Illustrations in the book are by one or other of the contributors, unless otherwise stated.

The contributors to this volume have been greatly assisted by reference to the following two publications:

Ronald Cox and Michael Gould. *Civil Engineering Heritage: Ireland*, Thomas Telford Publishing, London, 1998; and Ronald Cox (Editor). *Engineering in Ireland*. The Collins Press, Cork, 2006.

The publication of this book would not have been possible without the support and sponsorship of the Institution of Civil Engineers and the Electricity Supply Board.

RONALD COX AND PHILIP DONALD

CONVERSION FACTORS

Imperial measurements have generally been adopted for the dimensions of the works described, as this system was used when the great majority of them were designed and built. However, in 1970 the Irish construction industry adopted the metric system of units, so metric units have generally been used in relation to projects completed since that date.

The following conversion factors may be of use:

Length	1 inch = 25.4 millimetres	
	1 foot = 0.3048 metre	1 metre = 3.28 feet
	1 yard = 0.9144 metre	1 metre = 1.09 yards
	1 mile = 1.609 kilometres	1 kilometre = 0.622 mile
Area	1 square inch = 645.2 square millimetres	
	1 square foot = 0.0929 square metre	1 square metre = 10.76 square feet
	1 acre = 0.4047 hectare	1 hectare = 2.47 acres
	1 square mile = 259 hectares	
Volume	1 gallon = 4.546 litres	1 litre = 0.22 gallons
	1 million gallons = 4546 cubic metres	1 cubic metre = 220 gallons
	1 cubic yard = 0.7646 cubic metre	
Mass	1 lb = 0.4536 kilogram	1 kilogram = 2.20 lb
	1 Imperial ton = 1.016 tonnes	1 tonne = 0.984 tons
Power	1 horsepower (h.p.) = 0.7457 kilowatt	
Pressure	1 lb force per square inch = 0.06895 bar	
	1 lb force per square inch = 0.00703 Newtons per square millimetre	

INTRODUCTION

During the past three centuries, civil engineers have played a central role in providing the infrastructure essential to supporting civilisation on the island of Ireland. In particular, they have been involved in the provision of transport facilities, including roads, inland navigation, railways, docks and harbours, and airports; in the provision of water supplies, facilities for wastewater treatment and disposal, arterial drainage, and power supplies for industrial and domestic purposes.

Apart from a relatively short period of intensive industrialisation in the northeast of the island, centred mainly around the linen and shipbuilding industries, Ireland, unlike Britain, was never heavily industrialised. It has been estimated that in Tudor times the populations of Ireland and England were comparable, and much small-scale iron smelting was undertaken, using charcoal obtained from the indigenous oak woods. However, due to the effects of successive glacial periods in geological history, Ireland lacks the wealth of minerals found in Britain, in particular significant supplies of good quality coal. As a consequence, Ireland generally stood still in terms of industrial development and the need for major civil engineering works was thus correspondingly less.

The island of Ireland has a land area of 32,599 sq. miles (84,431 sq.km). Its greatest length is 302 miles (486km), its greatest width 189 miles (304km), and its coastline (depending on the method of measurement) extends for around 2,000 miles (3,200km). The island consists of an undulating central plain of limestone, almost completely encircled by a coastal belt of highlands of varying geological structure. The central plain is extensively covered with peat bogs and glacial deposits of sand and clay; numerous lakes dot its surface. The River Shannon, the longest river in these islands, drains this area, and has a basin equal to more than 20 per cent of the area of the entire island.

While much of the peat cover has been removed for fuel (including its use for the generation of electricity), the central Bog of Allen and other extensive areas of deep bog still form significant barriers to good communication. Eighteenth-century engineers overseeing the construction of, for example, the Grand Canal had to develop techniques for overcoming the inherent instability of these bogs.

The rivers of Ireland are generally small, although many discharge into deep estuaries. The River Shannon is a notable exception. Whilst its north–south course provided a useful waterway, it formed a distinct barrier to east–west communications. Most river systems have over the years been subjected, to a greater or lesser extent, to arterial drainage schemes, and many old stone bridges have required intervention to protect their foundations as riverbeds were deepened.

The population of Ireland was to increase to a peak of some eight million by 1830, but a series of famines, most notably the Great Famine period of 1845–48, led to many deaths and a significant increase in emigration. The land to the west is generally of poor quality and is often subject to adverse weather conditions emanating from the Atlantic Ocean. From the middle of the nineteenth century, the overall population started to fall, a decline only reversed towards the end of the twentieth century. The present population of the island is around 6.2 million.

Ireland's infrastructure has not only been subject to the same periods of unrest as in Britain, but, in addition, suffered the destructive effects of the Elizabethan and Williamite wars. Each of these episodes was followed by a series of Plantations, with land being given to settlers from England and Scotland. Many of the settlers brought with them their own construction techniques and expertise. One result of this periodic warfare is that there are very few civil engineering works in Ireland dating from before about 1700.

For a period in the eighteenth century Ireland had its own parliament based in Dublin. The Irish parliament was very supportive of industries, such as linen, and of certain civil engineering works, especially canals. Following the Act of Union in 1800, power was transferred back to London. The Westminster parliament is sometimes considered to have been less favourable to Ireland, but the establishment in 1832 of a Public Works Loan Board, able to give grants and/or loans, was to provide an important spur to the development of Ireland's infrastructure. Additionally, government support of public works was often seen as a way of assisting the unemployed poor.

Local government administration was slow in coming to rural Ireland. From 1710 Grand Juries were enabled to raise funds by 'presentment' for a limited range of works – most notably, from the civil engineering point of view, for roads and bridges. Later, guarantees were also given to support the extension of the railway system. Until the advent of public works loans, moneys for these works had to be raised from the larger landowners and progress was slow. Each Grand Jury had to decide its own policy in such matters, but pressure for road improvement also came from the Irish Post Office and most counties in Ireland undertook major road and bridge improvements in the nineteenth century.

From about 1880, Ireland was subjected to periods of agitation, which resulted in 1921 in the formation of the Irish Free State (Republic of Ireland from 1948). The new state covered twenty-six of the thirty-two counties, with the six forming Northern Ireland remaining within the United Kingdom. In this book, the term 'Ireland' generally refers to the whole of the island. Where a distinction is required, 'Republic of Ireland' and 'Northern Ireland' are used to signify the respective jurisdictions.

County Surveyorships were established in 1834, and in 1899 the civil engineering functions of the Grand Juries passed to newly established county councils. These bodies continued after the establishment of the separate jurisdictions in 1921, although in 1974 the county councils in Northern Ireland were abolished, the functions passing to government departments.

In Northern Ireland the protection of buildings of historical and architectural importance by statutory listing is well established. Following the practice in Britain, listing under the planning legislation was introduced in 1969. In certain circumstances, a partial grant is available to assist with maintenance and restoration. Under Northern Ireland legislation, the definition of 'building' includes any complete structure. Some 300 bridges of all types are listed, as well as other examples of civil engineering structures.

In the Republic of Ireland, the Architectural Heritage (National Inventory) and Historic Monuments (Miscellaneous Provisions) Act 1999 formally established the National Inventory of Architectural Heritage (NIAH) and placed it on a statutory basis. The NIAH provides planning authorities with information, which assists them in assessing and protecting the architectural heritage. This by definition includes structures associated with inland waterways and other civil engineering structures, especially bridges.

There have been a number of new planning acts – the first in 2000, amended in 2002, the most recent being the Planning and Development (Strategic Infrastructure) Act 2006. The Planning and Development Act 2000 – Part IV Architectural Heritage, deals with protected structures and architectural conservation areas (ACAs). The Department of the Environment, Heritage and Local Government (DoEHL) produced architectural heritage protection guidelines for structures of special interest and for the preservation of the character of ACAs. Under the Development Plans (Part II of the 2000 Act), each local authority is obliged to prepare a development plan indicating the objectives for its area, and this must include a Record of Protected Structures (RPS).

Successive chapters of this book deal with the heritage aspects of transportation infrastructure, such as roads, bridges, canals, inland navigations and railways. Other chapters discuss the civil engineering aspects of water supply, drainage and wastewater disposal, hydropower, and maritime facilities, such as docks, harbours, and lighthouses. The names of sites printed in **bold** also appear in the Gazetteer section of the book. This contains 156 entries arranged geographically and grouped according to province: Leinster, Munster, Connaught, and Ulster, in addition to Dublin and Belfast. Some of the entries include details of more than one site and the national grid reference and location is provided for each.

Where engineers are named, their dates are given in the index of persons, biographical information being generally available in one or other of the volumes of the *Biographical Dictionary of Civil Engineers of Great Britain and Ireland* published by Thomas Telford Publications, London.

There is a wide variety of civil engineering heritage to be seen in Ireland and it is hoped that this book will provide the information necessary to encourage readers to seek out and appreciate this legacy.

ROADS AND BRIDGES

Cable-stayed bridge carrying M1 motorway across the Boyne Valley near Drogheda, County Meath
ROUGHAN & O'DONOVAN/NARROWCAST

The cities and towns of Ireland are currently struggling to accommodate 21st-century traffic in unsuitable street layouts. Beyond the urban areas there are two road systems: recent dual carriageway and motorway layouts that have been superimposed on to a network of single-carriageway roads that evolved over several hundred years. In ancient times, between 509 BC (the beginning of the Roman Republic) and 476 AD (the collapse of the western Empire), the Romans built an empire that included all the coast around the Mediterranean, much of the Middle East, plus Spain and France, and which stretched as far as Hadrian's Wall in the north of England. To control their territories the Romans realised that communications and the ability to move men quickly would be the key. They also invented cement. Thus, with an unlimited availability of labour, they were able to build bridges, forts and roads all across their lands. Many of their routes are still in use today and many of their bridges still stand. However, the Romans did not cross the Irish Sea, and so no Roman network of roads was imposed on Ireland.

EARLIEST IRISH ROADS

The earliest road discovered in Ireland dates from 148 BC. It ran across a bog near Corlea, Keenagh in County Longford and consisted of pairs of longitudinal birch logs with transverse oak planks on top. The ends of the planks were pierced and birch pegs driven through that kept them in position. An 80m length of this road has been preserved by constructing over it a visitors' centre, which has controlled temperature and humidity to prevent the timber from deteriorating. The direction of the road indicated a route towards Lanesborough at the northern end of Lough Ree on the River Shannon, where there was a ford.

Written reference to a road system occurs in the *Annals of the Four Masters*, compiled by four friars in the 1630s from ancient writings, wherein it was found that in 123 AD roads wide enough to allow two chariots to pass each other were classified as '*slighes*'. There were five main *slighes* connecting ports, monasteries, towns and ecclesiastical centres. They radiated from the Hill of Tara and were *Slighe Asail*, which went west, *Slighe Cualann*, which ran southeast to Dublin, *Slighe Dala*, which ran south to County Kilkenny, *Slighe Mar*, (the Great Highway) which went northwest to Galway and the longest, *Slighe Midhluachra*, which ran due north to Antrim and then northwest to Derry, with spurs to Armagh and Downpatrick.

No information exists regarding road construction from that era and we can only assume that the roads, such as they were, were cut through forest, went around bogs, led to fords or short timber bridges and tried to keep to dry ground.

THE NORMANS, HENRY II AND THE TUDORS

In 1169 AD Diarmaid Mac Murchadha, Duke of Leinster, had his lands usurped by the High King of Ireland. Diarmud sought help from Henry II of England, who at that time controlled Britain and, through marriage, considerable areas of France. Henry sent Strongbow to Ireland and followed himself in 1171. (Diarmaid is still considered by many as a traitor in that he instigated 800 years of English rule in Ireland.) The Normans brought with them a new concept for roads: the King's highway, 'which is always open and which no man may shut by any threats, as leading to a city, port or town'. If this highway was locally closed by natural means, such as flooding or fallen trees, then the traveller could take the nearest available route without being accused of trespass.

By the mid-thirteenth century the Anglo Normans controlled Munster, the east coast and part of the Midlands, indicating that there had been a reasonable road network when they had arrived. Market centres developed into towns. An Act of 1285 passed in England but also applicable to Ireland stated that 'highways from one market town to another shall be enlarged, so that there be neither dyke, tree or bush, whereby man may lurke to do hurt, within two hundred feet either side'. It is doubtful if this was widely implemented because it is recorded that in 1558 it took two days for Thomas Radclyffe, an Earl and the Lord Deputy of Ireland, to travel the 60 miles from Limerick to Galway, and in 1602 it also took two days to travel the 30 miles from Newry to Downpatrick.

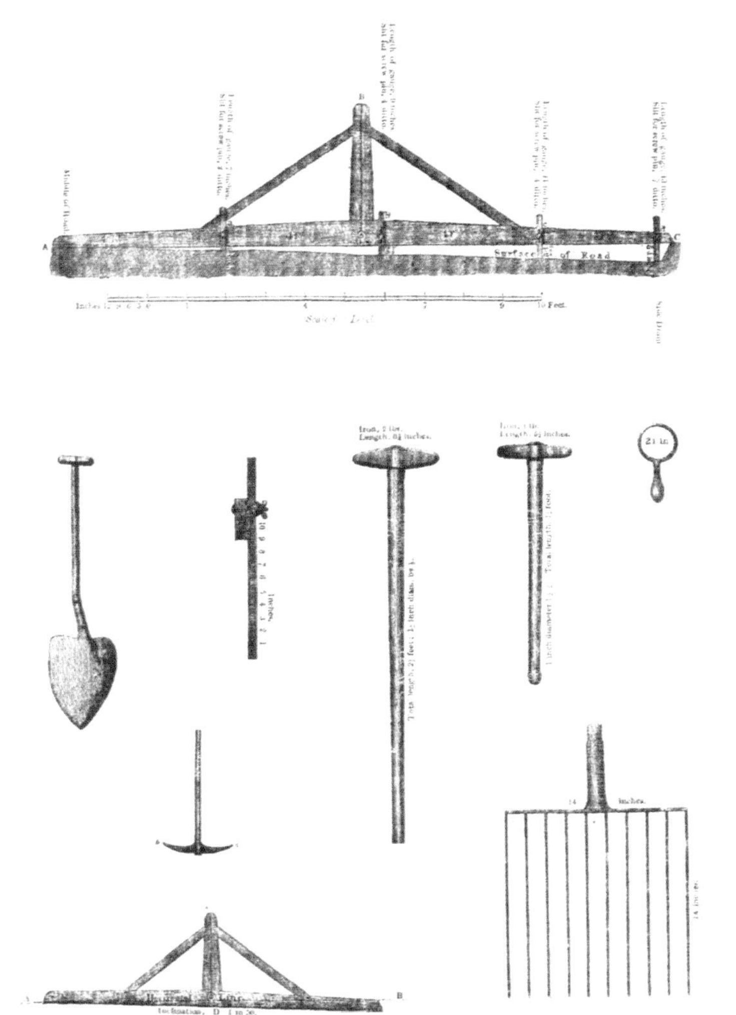

Early road-making tools and instruments

THE HIGHWAY ACT OF 1613 AND THE GRAND JURIES

In 1613 an Act was passed in the Irish parliament which encompassed the English Acts of 1555, 1563 and 1576. As a result, each parish was required to elect two people each year to act as Surveyors and Orderers for one year and they were to be responsible for organising the maintenance of the roads in the parish. The system became known as the 'Six Days' Labour' because at Easter each year six days were 'chosen for repairing and amending of highways and cashes and cutting and clearing of paces' for which the parishioners had to supply the necessary labour and materials. The 1613 Act did not authorise the construction of new roads but merely sought to keep existing roads in reasonable repair. Neither did the Act include for maintenance work to bridges but this was rectified by a further Act in 1634.

The Act of 1634 authorised 'the repairing and amending of Bridges, Causeyes and Toghers in the High-wayes'. The Act gave the Grand Juries powers to assess presentments and decide whether or not to proceed. Grand Juries were elected for each county plus Dublin city and generally included prominent citizens and landowners. Their primary function was to assess whether a prosecution case contained enough evidence to warrant the case to proceed to trial. The 1634 Act also asked them to assess proposals (presentments) for new roads, and decide if they should proceed and be constructed using funds previously allocated for proposed road construction. The Act stipulated that the new roads must be 30ft wide between ditches and that the central 14ft be paved with stones and gravel, materials which were usually readily available. Damage to the roads was minimised by imposing fines on wheeled vehicles that had wheels less than 3in wide. The idea for the system is attributable to Arthur French Esq. of Monivea, County Galway. In 1710 a further Act increased the powers of the Grand Juries to allocate funds to the repair of roads where the 'Six Days' Labour' system could not cope. In due course, in 1765, the 'Six Days' Labour' system was abolished.

All in all, the Grand Jury and Presentment system continued to work well for many years. It was a system quite unknown in England and excited favourable comment from English travellers. The most widely known of these was Arthur Young, a farmer who became a writer on matters agricultural and then wrote of his travels in Scotland and Ireland in the late 1770s. Regarding Ireland he wrote:

> For a country so very far behind us as Ireland, to have got suddenly so much the start of us in the article of roads, is a spectacle that cannot fail to strike the English traveller exceedingly. Any person wishing to make a road has it measured by two persons who swear to the measurement before a justice of the peace. It is described as leading from one market town to another, that it will be a public good, and that it will require such a sum to make the same. A certificate to this purpose is prepared and given to some

> one of the Grand Jury, usually in the spring. When all the common business of trials is over, the jury meets on that of roads; the chairman reads the certificate, and they are all put to the vote. This vote enables the person, who applied for the presentment, immediately to construct the road, which he must do at his own expence; he must finish it by the following assizes, when he is to send a certificate which is signed by the foreman, who also signs an order on the treasurer of the county to pay him, which is done immediately. The expence of these works is raised by a tax on the lands, generally between threepence and sixpence per acre. The effect of it (the Act) is so great that everywhere I found beautiful roads without break or hindrance, to enable me to realize my design. In a few years there will not be a piece of bad road, except turnpikes, in all Ireland.

Flowery language, probably somewhat exaggerated in its praise, but a worthwhile opinion none the less. Young also noted that Irish roads, once built, did not deteriorate as quickly as English roads, principally due to the type of traffic. In England there were many heavy wagons hauled by teams of four or eight horses. Of Ireland Young wrote, 'all land-carriage in that kingdom is performed with one-horse cars or carts', which meant that roads required much less maintenance.

In 1783 two English surveyors, George Taylor and Andrew Skinner, produced a book entitled *Map of the Roads of Ireland* which gave strip plans encompassing 8,000 miles of roads covering most of Ireland apart from the west, where there were very few roads. The strips did not require a countrywide survey triangulation and as such could not all be joined together to provide a map of Ireland. However, their 288 pages were of great use to travellers, showing as they did, distances in miles, churches, inns, racecourses, schools, army barracks, adjacent high ground, bogs and woods. These maps effectively give a snapshot of Irish roads at that time and a perusal of them shows that most of the 1780s network of country roads is still in use.

TURNPIKES AND THE POST OFFICE

In 1729 the first Act relating to turnpikes was passed. The turnpikes were funded by central government, tolls were charged and they were intended to complement the 'Six Days' Labour' and Grand Jury Presentment systems on the main routes. However, as Young noted above, he was not impressed by the quality of the turnpike roads. Problems had arisen because the tolls collected were absorbed by administration and the cost of collection, so that there was no money being ploughed back into maintenance.

In 1784 an Act established an independent Irish Post Office and it began to use the turnpikes and to pay tolls. An Act in 1792 enabled the Postmaster General (PMG) to give financial assistance to the turnpikes and the quality of these roads improved. Under a further Act in 1805 the PMG was able to set out basic design criteria in that gradients were not to be more than 1 in 35 and the width between fences was to be between 35ft and 40ft. Between 1805 and 1811, 1,988 miles of turnpike roads were surveyed; by 1818 the turnpikes were in good condition, and by 1829 were considered to be better than the Grand Jury roads.

In Munster the turnpikes received a further financial boost when an Italian immigrant named Carlo Bianconi discovered that after the Battle of Waterloo in 1815 the army had a surplus of strong, healthy horses for sale, and a year's supply of meal with which to feed them. Bianconi bought up these army horses and, based in Clonmel, started a passenger and freight business. In 1816 he was covering 226 miles a day and by 1843 he had 900 horses, 100 vehicles and was covering 4,000 miles a day in 22 counties. He was awarded contracts by the Post Office and was paying significant sums in road tolls for the use of the turnpikes. When the railways began to impinge on his business he moved with the times and established new routes that conveyed passengers and freight to and from the main rail centres. He became very widely known in business and local government circles and his carriages became known as 'Bians'.

Some new turnpike roads were notable, particularly the road from Tralee to Cork (now the N22), which enabled the farmers of County Kerry to get their butter more quickly to market in Cork. Considerable lengths of the original road are still identifiable, complete with side drains and boundary walls. Also, in County Down, just north of Dromore on the A1, a turnpike toll cottage still exists at the side of the twentieth-century dual carriageway.

DEVELOPMENTS DURING THE NINETEENTH CENTURY

In 1822 grants were awarded for the construction of roads to open up the western areas of the country where previously the road system had been virtually non-existent. In the southwest, particularly in the counties of Cork, Kerry, Tipperary and Limerick, the eminent

Turnpike Toll Cottage beside the A1 in County Down

engineers Richard Griffith, Alexander Nimmo and John Killaly designed and built roads between 1822 and 1840.

Whilst canals were being used for the transport of bulk materials, roads were increasingly being used for passengers and the transport of more valuable goods, such as linen. To facilitate carriages, interest was growing in actual road construction in relation to gradients, alignment, widths and the materials used in the road. Roman roads had a bed of stones about 8in deep as a foundation, which also acted as a drainage layer. On top were layers of stone of decreasing size, finished with a fine gravel as a running surface. The Roman methods were wasteful of materials so various more economic proposals were put forward, notably by Tresaguet in France (1775), Edgeworth in Ireland (1813), and Macadam (1815) and Telford (1820) in Britain. All their cross-sections exhibited the same method of construction with stones graded from large at the bottom to smaller at the top. It is felt that Edgeworth, in Ireland, was never given the credit he deserved as the forerunner of Macadam and Telford. Edgeworth used rounded stones with a blinding layer on top with a camber to dissipate rainfall. He also experimented with an improved suspension system for carriages to distribute the loads to his road surfaces better. Macadam and Telford advocated broken stone that would bind together better than rounded river gravel. Modern roads use the best of all these systems, complemented by better running surfaces.

In 1831 the Irish Board of Public Works was set up by the Westminster parliament. This Board amalgamated the various bodies concerned with fisheries, drainage, canals, poor relief and roads. The first new major road under the Board's jurisdiction was the 33-mile-long **Antrim Coast Road (U1)**, designed by county surveyor William Bald, which was a formidable civil engineering achievement combining, as it did, major rock excavations, viaducts, reclamation from the sea and drainage of a major bog. Generally, for the roads, the Board allocated funds to the Grand Juries whose duties now turned more to maintenance of the existing road layout rather than construction of new roads.

Antrim coast road (Limestone under Basalt)

From the 1830s onwards the railways began to make inroads into turnpike traffic and from 1,500 miles in 1820 the turnpikes had by 1831 shrunk to 780 miles and to 325 miles by 1856. In 1857 the turnpike trusts were wound up and their roads transferred to the county surveyors, who had been appointed from 1834. Finally, in 1898 the Grand Juries' road functions were handed over to newly formed county councils.

It is probably unwise to try and allocate road development to particular calendar periods, such as the nineteenth or twentieth centuries, because road development reacted to randomly occurring external events. From the time of the Napoleonic Wars and during the Victorian period, Britain was involved in wars and skirmishes all over the world, which meant, in effect, all over the British Empire. The event which would herald the end of the Empire and change the social structure of the western world was the so-called Great War of 1914–18.

AFTER THE FIRST WORLD WAR

During the First World War, mechanised vehicles incorporating the petrol-driven internal combustion engine came to the fore. For several decades transport by road and rail fought for supremacy, but slowly road transport took over from the railways in many countries, including Ireland.

In 1910 a Road Fund was set up which gathered income from taxes on vehicles and drivers. In the same year the First Irish Roads Congress was held at the Royal Dublin Society in Dublin, at which many topics were discussed, including the introduction of steamrolling of surfaces, concrete roads, and maintenance. For Irish roads the period from 1920 to 1960 was one of gradual improvement for the steadily increasing numbers of road vehicles. From the Road Act of 1920, main routes were established as trunk roads with link roads feeding in to them. With this classification the more important roads received higher grants. By the 1960s the use of bitumenised macadam (bitmac) and asphalt, both known as 'blacktop', was providing better and more durable running surfaces. (Asphalt was first laid in 1866 in Belfast by a Captain Shaw).

In 1922, the governments of Northern Ireland and the Irish Free State (or, as it became, the Republic of Ireland) went their separate ways. In the North, responsibility for roads devolved to different jurisdictions whereby Class I and II roads were looked after by the counties, unclassified roads by the rural districts and roads in towns by the urban districts. To facilitate the construction of new roads, legislation was passed in 1928 to enable land to be compulsorily purchased. Between the wars improvements were made to road surfaces by dressing them with tar in conjunction with the use of bitmac and the use of concrete. By 1935 there were 75 miles of concrete road, all of which has now been surfaced over. In Belfast the first dual carriageway was planned round Sydenham to the east of the city. The Second World War intervened and the bypass was not completed until 1958. Inevitably,

Interchange at Templepatrick, County Antrim on the M2 motorway (1960s)
UK MOTORWAYS ARCHIVE TRUST

after the war many of the towns on the trunk roads became increasingly incapable of handling the ever-larger commercial transport vehicles. Planning began for a motorway system that would have two lanes in each direction with very specific points for joining and exiting and with no right turns. The M1, west from Belfast, was built in the 1960s past Lisburn, Moira and Portadown as far as Dungannon. In the area to the south of Lough Neagh peat was removed to depths up to 40ft and the excavation backfilled with rock. The M2, going north, was completed in the 1960s and 1970s. It had a crawler lane added on the climb from Belfast up to Sandyknowes. For the stretch along the foreshore in Belfast reclamation was carried out by the use of 4 million tons of basalt overburden transported by rail from the Magheramorne cement works near Larne – the last use of bulk haul by rail and the last use of rail steam engines, ironically used for the construction of a new road. On the route south from Belfast various towns were bypassed by dual carriageways. As traffic still continued to increase, these dual carriageways, which permitted right turns and crossing, became increasingly dangerous and safety was improved by construction of overbridges and underpasses to separate through traffic from crossing traffic.

In 1973 responsibility for roads passed to the Department of the Environment, specifically to the Roads Service with headquarters in Belfast and six regional offices. In Belfast the Westlink dual carriageway was built in the 1980s. In the early 1990s the Cross-Harbour Bridges linked together the Sydenham bypass, the M2, Westlink and the M1. Early this century two roundabouts on the Westlink were under-passed and traffic now flows freely from central Belfast to Dublin via sections of safe dual carriageways and motorway.

In the Republic the Local Government Act of 1925 gave responsibility for roads to the county and urban district councils. Roads were classified as trunk roads or link roads. During the Second World War little more than basic maintenance took place. In the 1970s a national network of dual carriageways and motorways was planned, which began to be implemented in the 1980s. In 1993 a Roads Act was passed and the National Roads Authority (NRA) was established. During the 1990s and the first decade of the present century construction of the planned network continued apace and a very comprehensive motorway system now links all the major cities.

Most of the remaining towns that caused bottlenecks have been bypassed and three substantial tunnels built. The Dublin Port tunnel opened in late 2006. It was bored through from the M50 in north Dublin into the port area in order to take heavy vehicles off the

Breakthrough by Donegal miners in Dublin Port Tunnel DUBLIN CITY COUNCIL

surface roads. It is 4.5km long with two traffic lanes in each direction with headroom of 4.65m. In Cork an immersed tube tunnel under the River Lee was opened in 1999. (An immersed tube tunnel is built in large concrete sections in an area adjacent to the river. The sections which are each around 100m long are then floated out and sunk on to the river bed in a prepared trench where they are then joined together to form the tunnel). The Cork tunnel, which is 610m long with twin lanes in each direction, was named after a former Taoiseach, Jack Lynch (**Jack Lynch Tunnel (M28)**). In 2010 an immersed tube tunnel was also built under the River Shannon just to the west of Limerick city. It is 675m long, and like the Cork tunnel has twin carriageways. The Dublin and Limerick tunnels are tolled, although the Dublin Port tunnel is free for heavy goods vehicles (HGVs). With regard to roads in the Republic it could be said that we are returning to the days of the turnpikes, because several motorways carry tolls that are collected electronically or at toll booths.

Roads generally are required to cross many obstructions during the course of their length, such as rivers, streams, valleys, railways, and sometimes other roads. This part of the chapter considers the place of the bridge in Irish road networks, from those built of stone, to the more recent steel and concrete structures. Although timber was used in past times for many bridges, they will not be considered here as there are virtually no examples remaining due to them having become victims of Ireland's damp climate.

STONE BRIDGES

Road bridges built in Ireland prior to the middle of the nineteenth century are generally of stone, but are often referred to as masonry bridges, implying the skill of a stonemason. A cursory examination of a stone bridge will often indicate the level of skill involved, not only by the mason but, in more recent times, by the architect or engineer. It was not until the coming of the Cistercians (1148) and the Normans (1169) that the erection of stone arch structures of any significance began to be considered. The shape of the arch profile gives some indication of the age of a stone bridge, but must be taken in conjunction with many other factors. Semicircular arches were generally introduced into Ireland in the twelfth century and pointed segmental arch construction followed later. The three-centred arch is said to have originated in the sixteenth century. From it the false ellipse profile was developed that later became the standard for single-span canal and railway overbridges where headroom at the abutments was a critical factor. Flat segmental arches became common after 1750 when the balancing of the horizontal thrusting forces acting in the arch became better understood.

In order for an arch to remain stable, the line of thrust must lie within the middle third of the depth of the arch ring. If it passes outside the middle third, tensile stress develops, which may cause instability. During construction, however, for the process known as 'the turning of the arch', considerable temporary falsework is needed to support the weight of the material used to make up the arch. In brick construction, the arch may be built up from a number of rings of brickwork.

There are a number of ways in which an arch may fail, including development of excessive tensile or compressive stress in the jointing material, sliding of one voussoir (individual stone in an arch ring) over another, or spreading of the abutments. Other possibilities are failure of the foundations, fracture of the stone, or an arch may simply be overturned by the force of floodwaters. The jointing material used, depending on its nature, may offer some resistance to tensile stress. A good lime mortar will easily resist the tensile stress applied to it under normal conditions of loading. The condition of the fill above the arch ring also influences the overall strength of the structure.

Where it is considered necessary to extend the life of a masonry bridge on an important transportation route, major rehabilitation has often been undertaken. In many instances this has led to the spandrel walls being stabilised by being injected with liquid cement (grout) and the arch soffits sprayed with layers of liquid concrete (a process known as 'guniting'). This treatment alters the way in which the bridge was designed to act under load and tends to convert the multi-arch bridge into a monolithic structure little different to a solid causeway with openings. The original aesthetically pleasing outward appearance of many such rehabilitated structures has, however, often been retained.

Green's Bridge, Kilkenny

Notable early stone bridges include the fourteenth-century structures over the River Liffey at Kilcullen in County Kildare (1319) and at **Leighlinbridge (L2)** over the River Barrow in County Carlow (1320). Both have been much repaired and altered over time. In the sixteenth century, numbers of substantial stone bridges were erected, for example across the Boyne at Kilcarn near Navan in County Meath and in the larger towns of Ballinasloe, Carlow and Enniscorthy.

The vast majority of pre-1750 stone road bridges in Ireland are likely to have been planned and built by stonemasons, but, from around 1750 onwards, most large bridges were designed, and their construction supervised, by engineers, and occasionally by architects.

The Parisian architect and bridge engineer Jean-Rodolphe Perronet (1708–96) was the first to realise that the horizontal thrust of the arches was carried through the arch spans and that the piers carry only the vertical load and the difference between the thrusts of the adjacent arch spans. By keeping the span of each arch in a multi-span bridge the same there should theoretically be no thrust on the piers, so the piers could be greatly reduced in thickness. In order to maintain the stability of the piers during construction, the centring for all arches was erected and the arches built simultaneously, working from the piers towards the crown of the arch to minimise any thrust on the piers. The principal advantages were that the ratio of pier width to arch span could be significantly reduced, thus providing

Athlone Road Bridge

less of an obstruction to river flows, and the arch profiles could be made much flatter. Alexander Nimmo, designer of **Sarsfield Bridge (M9)** at Limerick, drew his inspiration from one of Perronet's most famous bridges, the Pont Neuilly over the River Seine in Paris (opened 1772). Nimmo's bridge is now of some historical importance following the replacement of the Parisian bridge in 1956.

In the 1760s, many of the road bridges spanning the rivers Nore and Barrow in County Kilkenny were destroyed or badly damaged by severe floods. George Smith designed replacement structures under an Inland Navigation Act of 1765. He was greatly influenced by the works of the Italian architect Andrea Palladio (1508–1580) and this can clearly be seen in the decorations applied to the piers and spandrels of a number of his bridges. In 1776, George Semple published his seminal *Treatise on Building in Water*, one of the classics of eighteenth-century civil engineering literature, which was to have a considerable influence on bridge foundation design and construction.

Only three masonry arch road bridges in Ireland have single spans equal to or greater than 100ft. A landmark in the construction of masonry arch bridges in Ireland was the bridge at **Lismore (M6)**, completed in 1775. It was not until 1794 that the Lismore span was exceeded in Ireland and then by only a few feet by Sarah's Bridge at Islandbridge in Dublin with a span of 104ft 5in. The last of this trio, and the largest masonry arch in Ireland is that erected in 1814 at **Lucan (D16)** with a span of 110ft.

Barrington Bridge (1818), County Limerick

The ability to span a river with a single arch, provided suitable foundations could be found for the abutments, reduced the risk of scour and the cost of construction and maintenance. However, the increase in span with an associated large rise in an arch created a new problem: that of excessive gradients on the approaches to the central arch or arches. This problem was overcome in large measure by the use of flatter segmental arches.

Bridges designed in the Gothic style with pointed arches in Ireland are rare, but that attributed to Alexander Nimmo at **Pollaphuca (L6)** in County Wicklow (1820) and that by Killaly near Lisdoonvarna, **Bealaclugga Bridge (M10)** in County Clare (1824) are worthy of mention. **Ringsend Bridge (D5)** in Dublin (1812), and the **Causeway Bridge (M7)** at Dungarvan in County Waterford (1816), of similar design, both have spandrel walls built as radial extensions of the voussoirs. The profile of the arch ring at Ringsend (like Sarsfield in Limerick) is hydraulically efficient and the masonry is continued across the bed of the river to protect the bridge against the scouring action of the river's flow.

As far as the dating of Ireland's considerable inheritance of masonry road bridges is concerned, the destruction of the bulk of the Grand Jury records in the Public Record Office during the Civil War in 1922 has left a void in information relating to bridges built and, even more importantly, repaired, in Ireland during the period 1700–1898. Style can act as a guide, but can rarely be relied on as sole evidence.

For many stone bridges, locally available stone or 'fieldstone' would have been used in their construction. This stone would have been naturally occurring and would have been either collected or quarried. Depending on the location, limestone, granite or sandstone would have been available, or some local geological variant of these rock types, such as 'greywacke'. Fieldstone was sometimes used in its naturally occurring state, otherwise the stone would have been dressed to fit as required. The stones would generally have only been roughly dressed. For major bridges where sufficient finance was available, good quality stone in large blocks could be quarried, often at some considerable distance from the bridge

Clara Vale Bridge, County Wicklow

site, or even from overseas. The stones were dressed by squaring off their faces, and these would be laid in courses. Stone, highly tooled to produce close-fitting blocks with very thin joints, is referred to as 'ashlar'. Joints in the stonework are generally filled with lime mortar.

IRON ROAD BRIDGES

Metal bridges, generally of iron or steel, are much less common in Ireland than in the industrialised areas of Britain and are mostly associated with railways.

The perfection of the process, whereby elements could be formed of cast iron in a mould, soon resulted in the construction of cast-iron bridges. As iron foundry technology advanced, bridge elements were soon being cast in larger sections and, by the time of the erection of the **Oak Park Bridge (L7)** in an estate near Carlow (erected 1818) the large imported castings forming the ribs of the bridge could be bolted together and cross-braced. The latter half of the nineteenth century saw the increased use of more ductile wrought iron rolled into plates and bars of differing sections.

There are only a few examples of wrought-iron truss road bridges in Ireland, those at **Ballyduff Upper (M25)** over the River Blackwater in County Waterford and the **Obelisk Bridge (L26)** spanning the River Boyne at Oldbridge being good examples.

CONCRETE ROAD BRIDGES

It was realised as early as the 1850s that iron bars inserted into concrete would make it stronger, but little was done initially to develop the idea due to the difficulty in obtaining bars of a sufficiently consistent strength. It is generally considered that the first practical

First use in Ireland of prestressed concrete beams (Sallins, County Kildare, 1952)

system of reinforced concrete was that developed in France by Monier (patented in 1867), which used two layers of small diameter round bars. It was the work of François Hennebique (1842–1921) after 1892 that led to the widespread use of reinforced concrete as a construction material. The granting of a licence by Hennebique to L.G. Mouchel in 1897 to design works in 'ferro-concrete' led to an explosion of concrete construction in Britain and, somewhat later, in Ireland. Mouchel undertook the design while construction was entrusted to licensed contractors, such as J. & R. Thompson of Belfast. Other designers, seeking to circumvent the various patents then in force, introduced alternative methods of reinforcement, often working on a design-and-build basis.

Resistance to shear was provided in ferro-concrete construction by flat bar stirrups, although round bar stirrups later became the norm in reinforced concrete. Many early bridges did not have the amount of shear reinforcement now considered essential and a number have been strengthened as traffic loadings have increased.

Precasting is the term used when sections or elements of a structure are cast away from the site and brought to the site for erection. Precasting allows work to be undertaken under much more even and controllable conditions. Bridges having precast deck beams or precast columns or balustrades are commonplace. As concrete is weak in tension, the carrying load on a beam may be increased if it is first put into compression. This is the underlying principle of prestressing. Wires, usually called tendons, are placed and then stretched before being anchored against the end of the beam. Release of the stressing-jack then introduces compression in the beam. Prestressed beams may be either pre- or post-tensioned. First used in 1952 in a road bridge replacement over the main rail line near Sallins in County Kildare, and, shortly thereafter in a bridge at Mount Norris in County Armagh, pre-tensioned beams are now widely used. The alternative method, and that needed for large beams, is to cast the beam first, leaving ducts for the tendons. These are then threaded in and stressed after the beam has attained its design strength. Large-span modern bridges are now frequently made up of a number of segments strung together in the manner of pearls on a necklace. The joints between each segment are glued using special epoxy resins, the beams subsequently being post-tensioned.

Most concrete bridges have a flat-deck profile, either with precast beams or cast *in situ* slabs, with or without attached beams. However, concrete arch bridges tend to look more pleasing and some exist with open spandrels. In the 1930s, a number of bridges with a bowstring profile were built, **Our Lady's Bridge (M27)** at Kenmare in County Kerry being a good example.

Large motorway interchanges, which incorporate several individual concrete bridges, appeared in Ireland with the opening of the first section of the M1 south of Belfast in 1962. The bridges on the early sections are of sturdy appearance with solid slab piers. As confidence in the use of concrete has grown amongst designers, the appearance of concrete bridges has become lighter with continuous beams and solid piers being replaced by open V-sections or rows of columns.

Balanced segmental construction has been used for large motorway bridges. This construction technique involves sections of cast concrete being erected alternately to either side of a support such that the out-of-balance force on the support is never more than that due to one segment. Match casting allows for more efficient jointing between the segments. During match casting a segment is formed in the casting yard. Its neighbour is then cast against the end, thus ensuring a perfect fit. This match casting process continues. The cast units are erected until one half of the structure has reached a little short of mid-span. When the corresponding half has been erected from the adjacent support, the two are formed into one continuous span by means of an *in situ* concrete 'stitch'.

Although concrete as a construction material is not without its problems, its cost-effectiveness and ability to span large gaps will ensure that it continues to be used for many bridge structures in the foreseeable future.

RECENT DEVELOPMENTS

Since the 1980s, there has been a significant development of Ireland's infrastructure, much of it resulting from an upsurge in economic activity and the provision of European Union and state funding, especially in the Republic. In 1993, the National Roads Authority (NRA) was established. New road building and improvements resulted in the rehabilitation or replacement of many bridges on existing routes and the erection of numbers of new bridges on, under or over motorways. In particular, the upgrading of that section of Euroroute 1 in Ireland, linking the ports of Larne in County Antrim with Rosslare Europort in County Wexford by way of Belfast and Dublin, has contributed to a large increase in bridge construction activity. The provision of new river crossings has resulted in a variety of design solutions and construction techniques being used in Ireland, many for the first time. Bypasses have been provided to remove heavy through-traffic from congested town centres and road alignments improved by the removal of sharp bends, often the result of the reluctance of early bridge builders to construct bridges on a skew.

Modern designers have the advantage of a better understanding of the properties of materials, better quality control of materials and the benefit of computers, which can now analyse more complex structures. Recent trends in bridge construction have been towards a greater use of precast elements. Recent bridge projects have generally favoured segmental concrete construction, such as the West Link Toll Bridge on the M50 in Dublin and the **Cross-Harbour Bridges (B2)** in Belfast, but there have also been some examples built in steel, mostly notably the **Foyle Bridge (U10)** over the River Foyle near Derry, and that at Wexford Harbour.

Research into the efficient design of precast concrete beams has resulted in the design of a range of standard types and their more frequent use. For example, a system of precast arch sections, 2.5m wide, for carrying roads over railways (or canals) on a marked skew, has been developed. In 1999, the NRA formally adopted the UK's Highway Agency's *Design Manual for Roads and Bridges*.

An awareness of the impact of the visual environment of bridges is growing and recent constructions have seen a marked improvement in bridge aesthetics. Improvements have come partly from more contracts being awarded on a design-and-build basis in open competition. This has provided an impetus for more innovative design, as has been evidenced by some recent bridge designs, such as the James Joyce and Samuel Beckett bridges spanning the River Liffey in Dublin.

Samuel Beckett Bridge, Dublin

The concept of a cable-stayed bridge is not new, but a recent upsurge in their use has resulted mainly from the fact that they offer the opportunity of crossing large and small obstacles with elegance and economy, the differing layouts of the cable-stays being one of the fundamental and distinguishing elements. The use of composite steel/concrete decks allows for a considerable reduction in the dead weight of such structures and in a simplification of their method of erection. Examples of large-scale cable-stayed bridges are the Boyne crossing near Drogheda and the recently erected bridge over the River Suir near Waterford.

Although bridges do not represent the bulk of expenditure on new road works, they do have a high visual impact on the public who use these transportation facilities. Motorway overbridges are frequently of a standard design, but attempts are now usually made to vary the detailing of piers and abutments to good effect, or to use the arch form to produce a more aesthetically pleasing structure, provided that a suitable foundation is available. Structural engineers and architects therefore put considerable effort into producing bridge designs that are both structurally sound and visually acceptable. From 1997 the pace of development increased and some hundreds of bridges have been added to the NRA's stock in the Republic to become an important and visible part of our civil engineering heritage.

CANALS AND INLAND NAVIGATIONS

Newry Ship Canal – Victoria Lock

In Ireland in the late seventeenth century there was discussion relating to the possibility of linking up Ireland's various rivers to improve communications and enable the bulk conveyance of goods. Eventually an Act was passed in the Irish parliament in 1715 'To Encourage the Draining and Improving of the Bogs and Unprofitable Low Grounds, and for the easing and despatching the Inland Carriage and Conveyance of Goods from one part to another within this Kingdom'. Various possible schemes were listed but as no funds were made available little happened.

From the map it can be seen that the network of canals and waterways, which were subsequently built in Ireland, evolved primarily from the east coast and extended westwards to Lough Neagh and the River Shannon. The first to be built was the Newry Canal (commenced in 1732), followed by the Tyrone Navigation (1733), the Boyne Navigation (1748), the Lagan Navigation (1753), the Shannon Navigation (1755), the Grand Canal (1756), the Barrow Line (1761), the Royal Canal (1789), the Ulster Canal (1830), the Ballinamore & Ballyconnell Canal (1842) and the Lower Bann Navigation (1847). The periods for construction varied from ten years (the Newry canal) to fifty-four years (the Tyrone navigation), so the canals came into use in a somewhat piecemeal fashion. As will be shown, the several canals met with varied commercial success and many, but fortunately not all, have now been closed to navigation. The rivers and canals run through attractive countryside and even where they are no longer operational, many still exist as channels to drain the adjacent countryside and can be seen from the fine masonry bridges that cross the navigations.

By way of clarification, a canal is completely man-made, whereas an inland navigation is a waterway which is part-canal, part-river. The river element needs to be slow moving with sufficient depth for barges. Locks enable canals to overcome changes in level. The principle of a lock was invented by the Chinese in the tenth century and refined by Leonardo da Vinci when he conceived of the idea of a pair of gates at each end of a lock closing in a 'V' pointing upstream. A canal that rises through a series of locks and which then has a level length before descending through further locks is known as a summit-level canal. As barges pass through the locks water is lost downstream. When a canal is part of a river system this loss of water is usually not a problem, but for a summit-level canal a constant supply of water is required. Other definitions relating to canals are given in the Glossary.

Generally using private money, the canals in Britain were built from the mid-eighteenth century for pragmatic and economic reasons: to move iron and coal to manufacturing centres and subsequently to move finished goods within the country and to the ports for export. Earlier, Ireland had also begun canal construction, but for a different reason. At the beginning of the eighteenth century Ireland was running out of its main source of fuel – wood, particularly in Dublin, where coal was being imported from England. Towards

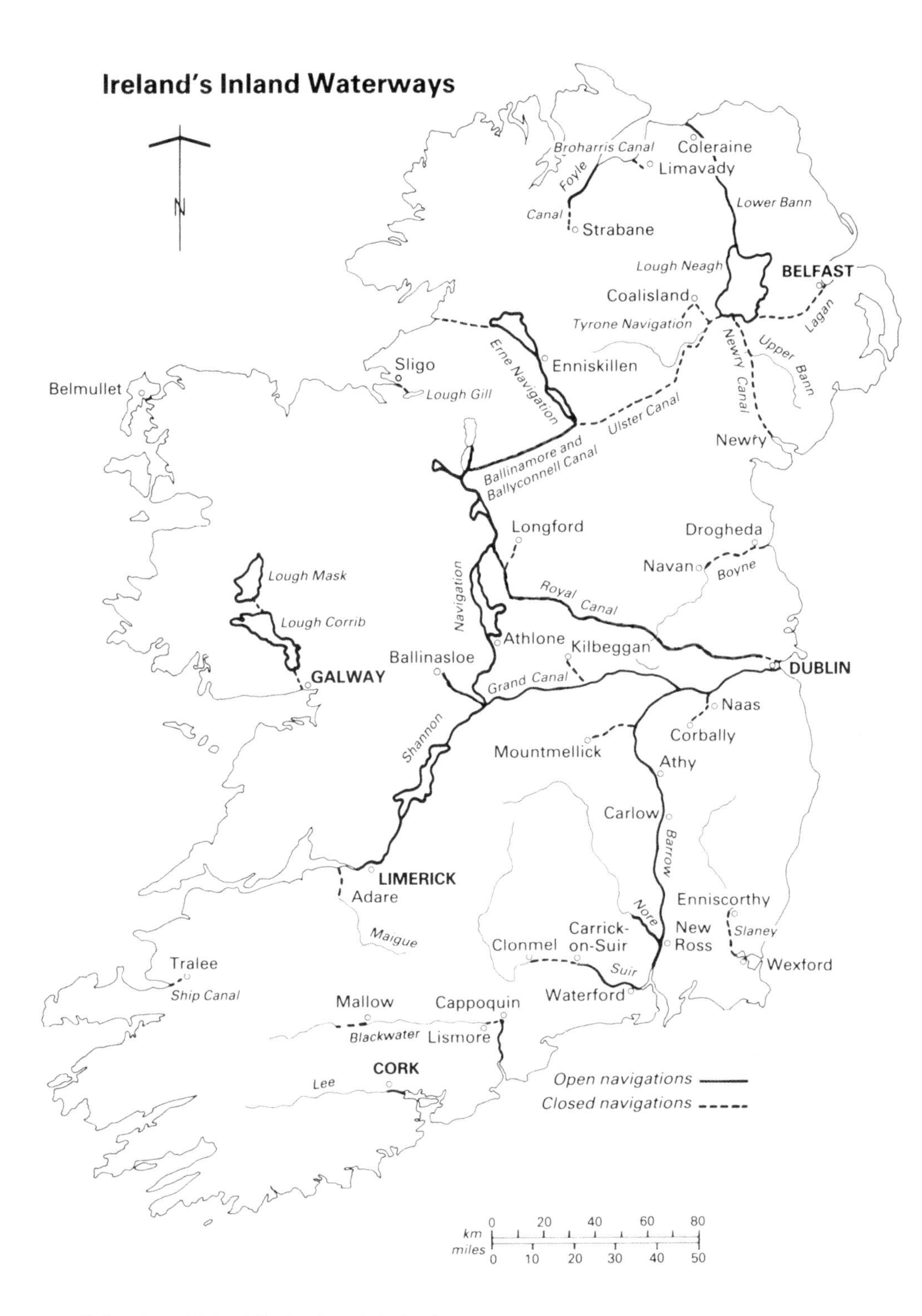

M2 Canals and Inland Navigations in Ireland RUTH DELANY

the end of the seventeenth century coal was discovered in outcrops on hillsides in east County Tyrone near Lough Neagh. Discussion in Dublin amongst interested merchants and politicians resulted in the Tillage Act of 1729 whereby funds would be raised by a tax on luxury goods, an eighteenth-century form of selective VAT. These funds would be used by the Commissioners of Inland Navigation for Ireland to finance the construction of an 18-mile long canal northwards from the port of Newry to join the Upper Bann river which flowed into Lough Neagh on its southern shore. From the southwest corner of Lough Neagh a complementary canal would be built from the River Blackwater to the Tyrone coalmines. Coal could then be transported economically from Tyrone via Lough Neagh and Newry to Dublin where it would be used primarily as a fuel and not for manufacturing purposes.

THE NEWRY NAVIGATION (U21)

Jerretspass Bridge over the Newry Canal in County Armagh

On Monday 2 April 1732 a labourer, employed for 7d per day and using his own spade, dug a sod of earth near Acton in County Armagh. Thus began the construction of the Newry Canal. It progressed well under the supervision of the Surveyor General, Edward Pearce, who was based in Dublin. With its eleven locks lifting the canal from Newry to Poyntzpass and the three locks lowering the canal from Scarva to the Upper Bann River, the Newry Canal, as well as being the first canal in Britain and Ireland, was also the first summit-level canal. Unfortunately, Pearce died in 1733. His assistant, Richard Castle (Cassels), took over until he was replaced by Thomas Steers in 1736. Steers had made his reputation building docks in Liverpool. Despite other commitments, Steers completed the Newry Canal in 1742 and a ship named *The Cope* laden with Tyrone coal sailed triumphantly into Dublin on 16 April. However, all was not as had been hoped. The coal had had to be brought overland to the banks of Lough Neagh and then transhipped for onward movement to Dublin, because the planned Tyrone navigation from the Blackwater to the coalfields, commenced in 1733, was foundering and would not be completed for many years.

Nevertheless, despite the lack of direct access to the coalfields, the Newry canal justified its existence by its transportation of local produce and general merchandise. In the latter half of the eighteenth century, 100,000 tons per annum was the norm and later, despite

Tandragee feeder aqueduct at Terryhoogan, County Armagh

the competition from the railways, traffic peaked at 127,000 tons in 1826. Initially, it was hoped that water from the adjacent Lough Shark would feed the summit level, but that was found to be inadequate and water was brought from the River Cusher via a small canal and a ten-span masonry aqueduct across a valley at Terryhoogan. To facilitate the passage of goods through Newry the canal was extended seawards by the construction of the Newry Ship Canal by Thomas Omer and his assistant, Christopher Myers, from 1759 to 1769. By the end of the eighteenth century the canal in general and the locks in particular were in a state of disrepair, because the lock walls had originally been built of brick and the floors of timber. Between 1801 and 1809, John Brownrigg repaired the canal and rebuilt the locks using cut masonry blocks. At Lock 11 near Poyntzpass the date 1801 is etched into a masonry block beside one of the timber lock gates. Brownrigg also rebuilt several bridges that crossed the canal. The most distinctive is near Jerretspass where the parapets uniquely rise to a point. Improvements by John Rennie were carried out to the Ship Canal between 1830 and 1850, and the basin in Newry was greatly enlarged and became the Albert Basin. Due to excessive water losses and a culvert collapse, further remedial works were implemented in the 1930s, which enabled ships up to 800 tons to reach the basin. The advent of the railways in the nineteenth century attracted trade away from the canal and the opening of the Newry–Armagh line in 1865, with the Belfast–Dublin line already in service, effectively marked the end of the Newry Canal as a viable enterprise. The last barge passed along in 1936 and the canal was closed in 1949.

Recently the Victoria Lock at the seaward end of the Ship Canal has been rebuilt (it was damaged by terrorist activity in 1969) and the canal is used by pleasure craft up to the Albert Basin in Newry. The 18-mile long towpath from Newry to Portadown has been reinstated and is part of the national cycle network, and a fine visitors' centre has been established in Scarva.

Restoration of 1st Lock on the Boyne Navigation in County Meath

THE BOYNE NAVIGATION

On the east coast the Boyne is an important river. In the 1740s Thomas Steers was asked to carry out a survey with the intention of making the river navigable to Navan, some 20 miles inland. Work began in 1748; Steers died in 1750, by which time the tidal lock at Oldbridge had been built. Omer continued in the 1750s and by 1766 the river was navigable for 9 miles as far as Slane using 3 miles of the river and four lateral canals with seven locks. Work was done upstream of Slane, but the works were liable to intense winter floods and were not opened. In 1787, 6,000 tons a year was passing from Drogheda to Slane. In the 1790s Richard Evans, assisted by Daniel Monks and John Brownrigg worked upstream from Slane using short lateral canals and completing the last 4½ miles by canal. New mills developed, but there was little other trade. There was debate about ownership of the navigation, there was siltation, and a general decline led to its closure in 1969.

Meanwhile, in Dublin and Belfast, by the mid-1750s, plans were being laid to build canals from Belfast to Lough Neagh and from Dublin to the River Shannon, as well as making the Shannon itself a navigable waterway. Why was all this being done? It was felt that the Newry Canal had been a success in that it had been built quickly, it was transporting considerable quantities of produce, and canals westwards from Dublin should meet with similar success. Also, there was still a lingering desire to connect all the main rivers of Ireland. So, in 1751, the Tillage Act of 1729 was extended for a further twenty-one years, to ensure that the revenue from that source would continue. Furthermore, from 1755, funds began to become available from the Irish parliament by a rather circuitous route. The English parliament had decreed that any surplus funds left at the end of each accounting period in Ireland would revert to the King. A group of members in the Irish parliament had no wish to finance the King and they undertook to ensure that there would be no surplus each year. They became known as the Undertakers and, until 1767, they allocated funds to finance the construction of the canal network, which satisfactorily mopped up any surpluses each year. Subsequently, some private funds were attracted and in 1800, just before the Act of Union, the Irish parliament appointed five Directors General of Inland Navigation and allocated them substantial funds. They were to control waterway development for the next thirty years.

THE LAGAN NAVIGATION (B7/U29)

In 1753, an Act was passed which authorised the construction of a waterway from Belfast to Lough Neagh. Under the supervision of Thomas Omer, construction commenced in 1756 at Stranmillis in Belfast, where a sea-lock was built on the River Lagan. The river continued to be used wherever possible with locks being built as required in short lengths of lateral canal. Lock 12 in Lisburn was reached in 1763. A further four years was required to reach Lock 13, which gave the canal a length to that point of 7 miles. The scheme then ran out of money. In 1768 an engineer, Robert Whitworth, recommended that work continue from Lisburn as a canal separate from the River Lagan. This proposal was adopted and a new company, the Lagan Navigation Co., was formed, which was controlled and largely financed by the Marquis of Donegall. Work recommenced in 1779 under Richard Owen (died 1830), who came from Flixton in Lancashire. He built two pairs of double locks (the Union locks) at Sprucefield and then continued with a summit-level length of 11 miles. The canal crossed the River Lagan by aqueduct and passed through the Broadwater, a small lake near Moira, en route to Aghalee. From there the canal descended through ten locks to arrive at Ellis's Gut on Lough Neagh in 1794.

The 27-mile long Lagan Navigation was commercially the most successful canal in the north of Ireland and contributed to the growth of the linen industry along its banks. In 1810 control of the navigation passed to Belfast merchants who carried out repairs and improved the towpath. From 40,000 tons per year in the 1830s, traffic increased to 170,000 tons a year by 1898, with 70 per cent of the cargos being coal moving upstream, which was the complete opposite of the original intention. The canal declined in the early years

Victoria Lock at Meelick, County Clare, on the Shannon Navigation, opened 1844

of the twentieth century and was closed in sections between 1954 and 1958, when the summit length between Sprucefield and Moira was infilled and used for the M1 motorway – an unparalleled act of heritage vandalism. The 300ft-long sandstone aqueduct across the River Lagan was also demolished.

With the advent of the railways, the station at Moira was sited close to the canal, where a superb example of a skew brick-arch bridge carries the railway across the canal. On the towpath at the corners of the bridge the metal guard posts can be seen, complete with the grooves worn by the towing ropes. In Lisburn, Lock 12 has been refurbished and is a feature of the Island Civic Centre, whilst near Belfast, Lock 3 has been refurbished with the adjacent lock-keeper's cottage opened to the public.

THE SHANNON NAVIGATION (C1/C2)

The Shannon waterway extends from Limerick for 160 miles and incorporates the longest river in Britain and Ireland, and its fifteen lakes. The waterway is divided into lower, middle and upper. The lower Shannon extends upstream from Limerick to Killaloe over a distance of 14 miles with a rise of 100ft. The middle section includes Lough Derg and the river then rises gently from Portumna via Meelick, Banagher and Shannonbridge to Athlone. The upper stretch includes Lough Ree and then Lanesborough, Tarmonbarry, Lough Forbes, Roosky, Jamestown, ending in Acres Lake in County Leitrim. Over the 146 miles of the middle and upper lengths the rise is only 60ft, which required a total of eight locks.

On the lower section from Limerick to Killaloe, work commenced under William Ockenden in 1757. He excavated a cut with locks to bypass the rapid fall in the river through Limerick. By 1759 he reported that completion was near. In 1767 the Limerick Navigation Company took over. Sixteen years later, after Ockenden's death, lateral cuts were still being excavated. In 1791 William Chapman presented his report, then joined the company and went on to complete a total of eleven locks. By 1799, barges were passing along the waterway to Lough Derg en route to Athlone.

Construction of a navigable waterway north from Portumna to Athlone began in 1755 under Thomas Omer and was complete as far upstream as Roosky by 1769. Omer built a lock at Meelick and bypassed Banagher with a lateral canal and by a flash lock. Omer also designed and built a bridge and lock-keepers' accommodation, and a bridge-master's house survives at the eastern end of the Shannonbridge in Banagher. In the 1770s, the short length of Jamestown Canal was built. After 1800 the Directors General took charge. At this time the Grand Canal was approaching from Dublin and their directors expressed concern regarding the state of disrepair of the Shannon Navigation. Their engineer, Richard Jessop, and Director General John Brownrigg decided what needed to be done and the necessary work was completed by 1810. Steamers were introduced to the waterway in 1826. They transported goods and passengers and towed barges, but there was little commercial activity above Athlone. Control passed to the Shannon Commissioners in 1831 and Thomas Rhodes, using steam-driven dredgers, improved the channel and, by 1850 at Athlone, he had built a weir and a bypass lock. Trade was around 100,000 tons per year but by the 1880s, with the advent of the railways, this had dropped to 70,000 tons. A slow decline continued, but nowadays the Shannon Navigation is a tourism success story, due mainly to the efforts of the Inland Waterways Association of Ireland, a group of enthusiasts who would not let the Shannon waterway die. Whilst the mix of river and lakes is attractive for tourists, the many lakes had been a handicap for the bargees in the eighteenth century. On a lake there cannot be any towpaths so traversing the lakes meant slow progress downstream on the flow and either poling or waiting for a favourable wind to make progress upstream.

In the late 1920s, a large hydroelectric scheme was built at Ardnacrusha above Limerick. Lough Derg was raised by some 6ft and a double lock with guillotine gates installed, which delivered boats down through the 100ft fall to the tailrace below the power station. These locks superseded all the existing locks.

THE GRAND CANAL (L8)

As a result of the 1715 and 1729 Acts various routes for a canal going west from Dublin had been suggested and surveyed until, in 1757, the Grand Canal was commenced to the west of Dublin under the supervision of Thomas Omer. He had proposed to take the canal straight through the Bog of Allen but John Smeaton suggested a route further to the north towards Edenderry. Smeaton's suggestion was adopted although it still meant some miles through the bog. In 1772 the Grand Canal Company was formed and under John Trail and Charles Tarrant the canal by 1779 stretched from St James's Street in Dublin to Sallins, 18 miles to the west. In 1773 Trail and Smeaton disagreed over the size of the locks and, as a result, Locks 11, 12 and 13 near Lucan are a compromise between Trail's large locks and Smeaton's recommended smaller ones. Smeaton's size prevailed for the remainder of the canal. A passenger service between Sallins and Dublin was inaugurated in 1780. The canal crossed the River Liffey on the **Leinster Aqueduct (L9)**, designed by Richard Evans with five spans of 25ft, and rose through nineteen locks to its summit level near Lowtown in County Kildare, 280ft above low water at Dublin. The summit was supplied with water from the Milltown and Blackwood feeders.

The Bog of Allen presented a problem that no engineer had ever had to face in England. Smeaton felt that the canal should be built at the same level as the bog, opining that one should 'avoid a bog if you can, but by all means possible, the going deep into it'. William Chapman disagreed, but Smeaton prevailed and his route through the northern edge of the Bog of Allen has given trouble from time to time, when breaches have occurred in the canal sides. Later, the method of cutting longitudinal and transverse drains and then allowing two years for drainage and subsidence proved to be more successful. On the Barrow Line, Bernard Mullins gained valuable experience. He subsequently became a partner in the successful contracting firm of Henry, Mullins and MacMahon. After their construction on the extended Grand Canal of the Ballinasloe spur, which was in bog most of the way, Mullins wrote a treatise on how to build roads, canals and railways across bogs.

Under Richard Evans and John Killaly, the Grand Canal reached Daingean, 15 miles before Tullamore, in 1797, and then Tullamore in 1798. By 1796, at the Dublin end, William Chapman had brought the canal in a loop, known as the Circular Line, from St James's Street in Dublin down through seven locks to the Grand Canal Dock.

From the summit at Tullamore a route to reach the Shannon was surveyed and, using the valley of the River Brosna, the Grand Canal reached Shannon Harbour above Banagher in 1804. The Napoleonic Wars increased trade, which rose to 100,000 tons per year in 1800 and 200,000 tons by 1810. Along the route five hotels were built that had varying degrees of commercial success. Between 1786 and 1835, spur canals were built to Naas and then on to Corbally, to Ballinasloe beyond the Shannon, and Kilbeggan, which helped growth in those towns and trade in general. On the Naas spur, Chapman devised a method for building masonry bridges on the skew across the canal. His three bridges were subsequently demolished, but two examples remain on the Grand – the Shee Bridge near Robertstown, and later the railway bridge over Ringsend Basin. Trade decreased after the Battle of Waterloo in 1815, surprisingly picked up again with the advent of the railways,

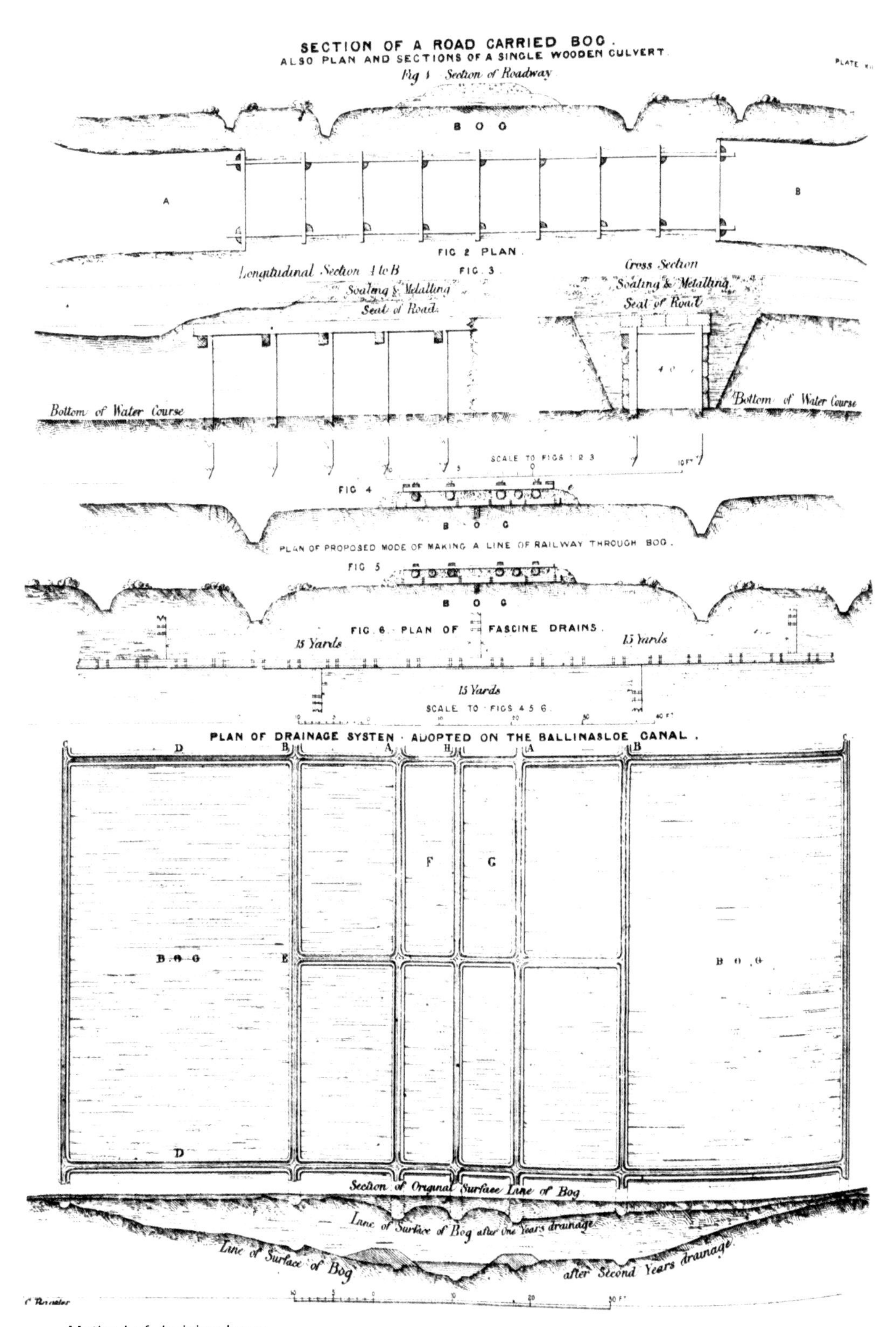

Method of draining bogs

Road lifting bridge on refurbished Royal Canal in County Longford

and then dropped off again briefly in the post-Famine years after 1848 and finally increased to 300,000 tons by the end of the 1850s. Competition from rail and road increased and the canal was sold to CIÉ in 1950. In recent years the canal has been refurbished and is now being used for tourism along its entire length.

THE BARROW LINE AND NAVIGATION (L10)

The River Barrow is navigable from St Mullins to the sea. In 1761 Thomas Omer began to work upstream from St Mullins, dredging and building lateral canals, with locks where necessary. Chapman took over in 1790 and reached Athy using the river and more lateral canals.

In 1783, before pushing on to the Shannon, the Grand Canal directors decided to branch off to the south to link up with the River Barrow. It was here that Mullins worked his way through the Ballyteague bog. By 1785 Monasterevin had been reached. There were many shallows to the south and, under Archibald Millar, a canal was built for 15 miles to Athy by 1791. The Barrow Line incorporated an aqueduct (**L11**) and some interesting bridges. The aqueduct brought the canal over the River Barrow near Monasterevin. It had three spans of 41½ft and was designed by John Killaly's son, Hamilton, and built in 1826. Nearby, to carry the road across the canal a lifting bascule bridge was built, which is known locally as the 'drawbridge'. A similar lifting bridge can be seen near Bagenalstown. Also, near Nevinstown, a footbridge, known as the 'guillotine' bridge, lifts vertically when headroom is required for passing canal traffic. (Similar bridges have recently been installed at a number of locations on the refurbished Royal Canal). From Monasterevin a spur was built to Mountmellick.

THE TYRONE NAVIGATION

In 1767, the Newry Canal had been completed for twenty-five years, the Lagan Navigation had reached Lisburn, the Shannon Navigation above Killaloe was well under way and the Grand Canal had commenced ten years earlier. However, the Tyrone Navigation, begun in 1733, was still not functioning. Acheson Johnston had been instructed to build a canal to the coalfields of County Tyrone from a point on the River Blackwater. He chose a place from which he would commence the canal and it has been known ever since as 'The Point'. Johnston had no building or architectural experience and the only thing he knew about canals was that they needed a water supply. The Point was adjacent to the River Torrent, which was very appropriately named. Johnston struggled through a bog alongside the Torrent for 3 miles and then struck off to the northwest to a reasonably level area at Coalisland. By 1767 he was still trying to complete the 7 miles of canal, and in the meantime it had been discovered that the richest coal seams were not in fact at Coalisland but at Drumglass, 3 miles further to the north, and 200ft higher.

To reach Drumglass from the basin in Coalisland various schemes were proposed. Christopher Myers suggested a 'standard' scheme incorporating numerous locks, but it would probably have been impossible to replenish them with adequate supplies of water. An architect named Davis Ducart then proposed to surmount the 200ft by a system of horizontal water-bearing tunnels connected by vertical shafts. (James Brindley had used a similar arrangement on his recently completed Bridgewater Canal). Ducart was from Sardinia and his anglicised name derived from Daviso de Arcort. Outside Dublin he had designed and built several fine country houses in the Palladian style and the Custom House in Limerick, which now houses the Hunt Museum. Money was allocated to Ducart's proposals and he began by building two short lengths of canal and an aqueduct across the River Torrent near Newmills. In building the three-span aqueduct, Ducart was doing what he knew best – building in masonry. He then applied for further funds and, when they were allocated, he changed his scheme. The tunnels and shafts were removed and replaced by four lengths of canal linked by three inclined planes, built of stone and earth, which were known locally as 'dry hurries'. **(U19)** He proposed that the coal would be transported in wheeled tubs on lighters on the lengths of the canal. The tubs, each carrying two tons of coal, would then be lowered by winch down the slope of the inclined plane to the next length of canal below. John Smeaton, who was in Dublin at the time, was consulted regarding the best way to raise and lower the tubs on the inclines. He sent his assistant,

Ducart's 'dry hurry' at Drumreagh in County Tyrone

William Jessop, and it was recommended to Ducart that the tubs be balanced as far as possible by having a descending full load pulling up an ascending empty tub. It is reported that coal progressed thus from Drumglass to Coalisland in 1777, but the scheme was found not to be viable due to the lack of suitable power to drive the winches. Ducart did not have sufficient water to harness for power and he was just too early for steam power. Ducart's canal was abandoned and the Coalisland Canal was eventually completed in 1787. Like the Newry Canal it needed repairs in the early nineteenth century and Thomas Monks rebuilt the locks between 1801 and 1809. Trade increased during the nineteenth century to 36,000 tons per year and peaked at 57,000 tons in 1931, but then fell away sharply. The canal was closed in 1952 and the basin in Coalisland was infilled in the late 1950s.

Little remains of Ducart's canal apart from the aqueduct, which is now a viaduct, together with elements of two of the inclined planes. The towpath alongside the Coalisland Canal has been refurbished and is used by local walkers.

THE ROYAL CANAL (L12)

Two directors of the Grand Canal Company broke away and incorporated the Royal Canal Company, which was fortunate to receive a grant at the time of a debenture scheme to encourage public works. It was suggested that the Royal should share the Grand Canal to Kinnegad, some 35 miles west of Dublin, before taking its own route to the northern

Dry dock on the Royal Canal at Richmond Harbour in County Longford

reaches of the River Shannon. The proposal was rejected, and in 1789 construction began on the basis of a survey by John Brownrigg with the details by Richard Evans. The Royal Canal began with a sea lock at the North Wall quays in Dublin, crossed the Rye Water near Leixlip, went through the Bog of Cappa, had reached Thomastown by 1805, climbed through eight locks to the summit, reached Mullingar in 1806 and the western end of the summit, 6 miles beyond Mullingar, by 1809. The summit was fed from nearby Lough Owel. Broome Bridge crosses the canal in Cabra in Dublin. It was here on 16 October 1843 that the celebrated mathematician William Rowan Hamilton was walking along the towpath when he had inspiration and on the bridge inscribed his formula relating to quaternions: $i^2 = j^2 = k^2 = ijk = -1$. A plaque on the bridge commemorates the event. The aqueduct, which takes the canal over the Rye Water, was not the normal masonry structure, but an

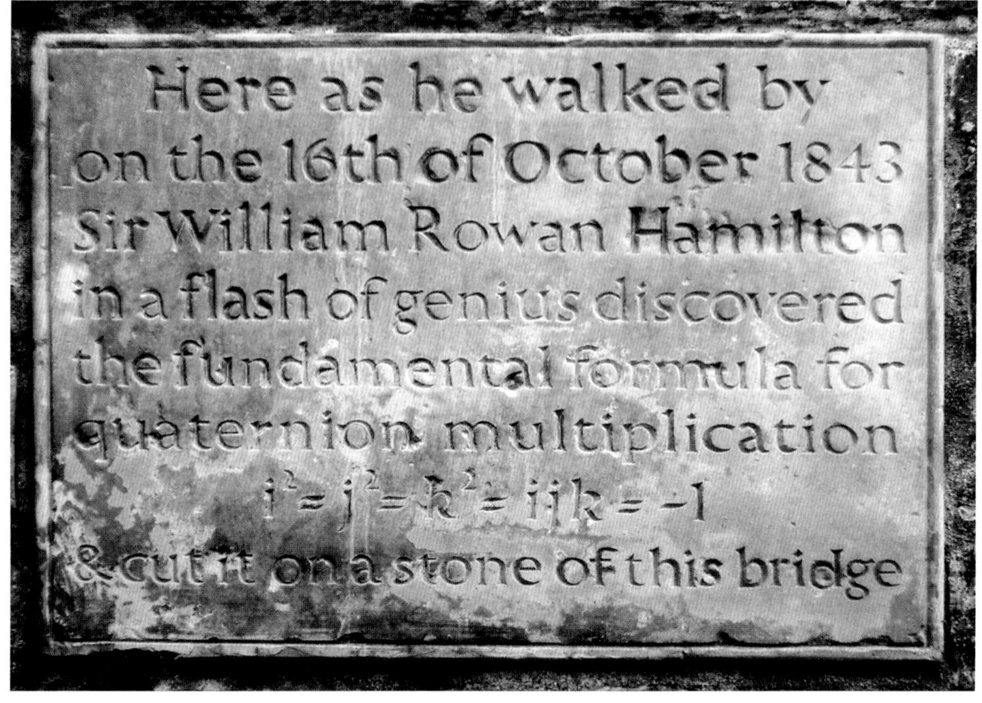

The Hamilton plaque on Broome Bridge over Royal Canal in Dublin

embankment 90ft high, which also carried the railway in later years. The river passed through the embankment via a masonry arch 'tunnel' 230ft long and 30ft wide.

There was then a bitter debate with the directors of the Grand Canal as to where the Royal should join the Shannon. The Royal directors wished it to end in Lough Ree, a few miles north of Athlone, but this was vigorously rejected by the directors of the Grand as they said the Royal would be so close to Athlone that it would affect their trade. In 1813 the Directors General took over, awarded the contract to Henry, Mullins & MacMahon and, with John Killaly in charge, the Royal joined the Shannon at Richmond Harbour near Cloondara in 1817. In its 90 miles the Royal has four aqueducts – from east to west they cross the M50, the River Blackwater, the River Boyne and the River Inny. Richard Evans designed the **Boyne Aqueduct (L13)** with three spans of 40ft, and later Killaly designed the **Whitworth Aqueduct (L14)** to take the canal over the Inny near Abbeyshrule. It has five spans of 20ft in a total length of 165ft. The canal incorporated twenty-five locks on the ascent to the summit level and twenty-one on the descent. Like the Grand Canal, spur canals were built, to Broadstone in Dublin and to Longford.

In 1810 trade was measured at 52,000 tons. In 1844 the Midland Great Western Railway bought the canal so that they could lay their track bed alongside. Trade on the

Whitworth Aqueduct carrying Royal Canal over River Inny in County Longford

canal declined and it was formally closed in 1961. In recent years the Office of Public Works and Waterways Ireland have restored the canal, initially from Dublin to beyond Mullingar, and subsequently, along its entire length to the Shannon Navigation.

THE LOWER BANN NAVIGATION (U7)

Several rivers flow into Lough Neagh, but the only outlet to the sea is via the Lower Bann from the northern end of the lough to Coleraine. For centuries, land around Lough Neagh flooded. When the canal era began it was suggested that a rocky outcrop near Portna could be removed to improve the flow in the Bann. John McMahon surveyed the river in 1840, an Act was passed in 1842, and work began in 1847 under Charles Ottley. Six locks were installed between Toome and Coleraine and, while matters improved, complaints about flooding began again in 1859. Nothing was done until 1925 when Major Shephard arranged a resurvey. It was discovered that not all of McMahon's recommendations had been implemented, probably because he had died before its completion in the nineteenth century. In 1930, new sluice gates at three of the sites were installed, the level of Lough Neagh was lowered, and the incidence of flooding diminished.

Lifting rail bridge over the River Bann near Coleraine, County Londonderry

Travelling up the River Bann from Barmouth near Castlerock, the first bridge is the rail bridge on the Coleraine–Derry line (**U8**), which has a lifting span designed by Strauss, who went on to build the Golden Gate Bridge in San Francisco. The first lock is at the Cutts, above Coleraine, where it behoves the sailor to take notice of warnings that floodwaters from Lough Neagh may be approaching. At Drumaheglis near Ballymoney there are moorings, a slipway and a caravan site. There is also a marina at Portglenone. The two locks at Portna are in a parallel length of canal, and above Toome the final lock brings the Lower Bann to its source at Lough Neagh. In bygone days this would have given access to Belfast via the Lagan Navigation; to the Newry Canal via the Upper Bann; to the Ulster Canal and the Tyrone Navigation via the River Blackwater.

THE ULSTER CANAL

The ambition to link Belfast with Limerick was to have been fulfilled with the construction of the Ulster Canal from Lough Neagh via the River Blackwater to the Erne, and the Ballinamore & Ballyconnell Canal from Lough Erne to the northern end of the Shannon Navigation. In 1815 John Killaly surveyed a route for the Ulster Canal and in 1825 proposed a scheme, which included eighteen locks. The Irish Loan Commissioners consulted Thomas Telford and the number of locks was increased in 1831 to twenty-six. William Dargan was awarded the contract. From Charlemont on the River Blackwater, the summit level near Monaghan was reached in 1838 via a total of nineteen locks and, by descending through seven locks, Wattle Bridge on the Finn River at the southern end of Lough Erne was reached in 1841 after 46 miles. Incredibly, the final lock at Woodford was built with a width of 11ft 8½in, narrower than any other lock in Ireland, which meant that barges built for the Newry or Lagan canals could not pass along the Ulster Canal, thus compromising the possibility of a through route between Belfast and Limerick. The reason for this width of lock has never been established, although several theories have been put forward. One of these suggested that if 'normal' barges were unable to use the canal then Dargan could put a fleet of his own narrow barges on the canal, which merchants would be forced to use, to his benefit. Regardless of the narrow lock issue, the Ulster Canal was never a success, because the length of the summit level could not be supplied satisfactorily with water from Quig Lough, near Monaghan. Remedial works were carried out from 1865–73, but the problems were not cured. Against their better wishes the Lagan Navigation Company purchased the Ulster Canal in 1888 and struggled with maintenance until the canal closed quietly in 1931. Subsequently, some lengths of the canal have been infilled, some locks plundered for their masonry blocks, and some small low-level bridges built. Even in its derelict state there are areas worthy of a visit – a lock and a restored mill adjacent to the stretch at Benburb Gorge, and the single-span Ardgonnell aqueduct near Middletown.

In recent years surveys have been carried out with a view to reopening the canal. It would require some excavation to avoid buildings encroaching on the original line, the construction of several bridges and locks to give headroom under the low-level bridges and the reconstruction of the narrow locks to be consistent with the locks on other canals.

THE BALLINAMORE & BALLYCONNELL NAVIGATION (C3)

The final link in the Belfast–Limerick route was the Ballinamore & Ballyconnell Navigation linking the Shannon Navigation to Lough Erne. Complications arose because the canal was funded by the Arterial Drainage Act of 1842 and it was to be a combined drainage/navigation system with the construction costs allocated to each. The scheme was designed by William Mulvany and John McMahon. The navigation started near Leitrim village and rose through eight locks for 3 miles to Lough Scur. There was a summit level for 3 miles to Lough Marrave, leaving 32½ miles to River Erne using the Woodford and Yellow rivers and eight locks. Due to over-expenditure, economies were forced and there were deficiencies in construction. Notwithstanding the difficulties, the navigation opened in 1860. In 1868, a steamer called the *Knockninny* took three weeks to pass along the system with water having to be moved from one reach to the next. In nine years only eight vessels used the navigation and it was closed in 1869.

Renamed the Shannon–Erne Waterway, this cross-border navigation was completely refurbished between 1990 and 1994 under a scheme jointly financed by the governments of the two jurisdictions in Ireland. Five locks were rebuilt, the gates were replaced on eleven others, 6 miles of the canalised section were deepened, bridges were refurbished

Shannon–Erne waterway

and now tourist traffic is able to travel enthusiastically from the Erne to Limerick via the Shannon, and from the Erne to Dublin via the Shannon and the refurbished Royal or Grand canals.

OTHER CANALS

Other canals and navigations were built at Strabane, the Foyle, Tralee, Lough Corrib, Belmullet and Lough Erne.

Strabane: Between 1791 and 1796, after taking advice from Richard Owen, the Marquis of Abercorn underwrote two-thirds of the construction costs of a 4-mile length of canal with two locks from the River Foyle to Strabane, to enhance trade to the extensive lands he owned in the area. As a result, Strabane prospered in the first half of the nineteenth century with traffic peaking at 10,000 tons in 1836. With the coming of the railway to the area in 1847 there was a slow decline in trade. By the 1930s there was no traffic and the canal was closed in 1962.

Foyle: From Lough Foyle a drainage/navigation canal was excavated in the 1820s for a distance of 2 miles towards Limavady. It was known as the Broharris Canal and helped local trade for a period, but it never reached Limavady.

Tralee: Between 1832 and 1846, a ship canal with one sea lock was built for 1½ miles to bring low-tonnage ships into Tralee. The canal prospered for forty years and for a period emigrants went to America, the ships returning with meal. The canal silted up and was closed in the 1930s when a deep-water quay was built at Fenit. In the 1990s and early 2000s, the sea lock and the canal were restored up to the basin at Tralee.

Eglinton and Cong Canals (C11): Various schemes to connect Lough Corrib with the sea were proposed until finally it was decided that improved trade would benefit an impoverished area. Under John McMahon a ¾-mile long canal with two locks was commenced in 1848 to bypass Galway to the west. It was opened by the Earl of Eglinton in 1852. Steamers operated on the lake in the 1850s and 1860s. The canal did not reopen to commercial traffic after the First World War. Low bridges built in Galway in the 1950s now prevent passage along the Eglinton canal. The Cong Canal between Loughs Corrib and Mask was also started in 1848 but was abandoned in the 1850s on the premise that the limestone rock would not hold water. It is more likely that work was stopped due to costs and the advent of the railway.

Facing page: Parkavera Lock on the Eglinton Canal in Galway

Belmullet: The Belmullet Canal, at a quarter of a mile long, is the shortest canal in Ireland. To avoid 15 miles of the Atlantic Ocean it was cut across the neck of the Belmullet Peninsula in the 1880s to give ships safe passage between Blacksod Bay and Broad Haven.

Lough Erne: The River Erne provides a natural 40-mile waterway between Belleek and Belturbet, Enniskillen being located halfway between Upper and Lower Lough Erne. In 1841, after the completion of the Ulster Canal, William Dargan brought a paddle steamer to the lakes, ostensibly for passengers, but in reality principally for trade. This pattern continued for the next 100 years. In 1957, for the cross-border hydroelectric scheme at Ballyshannon, a barrage and lock were installed downstream of Enniskillen to control the level of the Upper lake. The Lough Erne area is now linked to the Shannon via the Shannon–Erne Waterway.

CONCLUSIONS

The early canals were conceived for the right reasons, but it is difficult to justify the subsequent proliferation of canals in Ireland where there were no natural resources that would have led to industrialisation. The writer Arthur Young summed it up very presciently in the 1770s when he said 'have something to carry before you seek the means of carriage'. There was also a lack of overall control as locks varied considerably in size culminating in the Ulster Canal fiasco, despite a century of previous experience. However, on a more positive note, the construction of the canals employed considerable numbers of men in times of great unemployment (up to 3,000 men could be at work digging a cut at any given time), and the canals subsequently gave employment to bargees, lock-keepers, stable hands, boat-builders, joiners and masons. Also, the midlands of Ireland became more accessible, towns along the routes were enhanced and new towns sprang up round inland harbours at Scarva, Coalisland, Ballynacargy, Cloondara, Kilcock, Edenderry and Daingean.

While the canals and navigations were not a real commercial success, we are fortunate they were not completely lost. Due to strenuous efforts by individuals and government, the Shannon Navigation, the Grand Canal, the Ballinamore & Ballyconnell Canal, and most recently the Royal Canal, have been refurbished and are now major tourist assets. Work has also commenced to refurbish and eventually reopen the Newry Canal, and the Lagan Navigation between Belfast and Lisburn.

RAILWAYS AND VIADUCTS

Valentine's Glen rail viaducts in Greenisland, County Antrim

Ireland's railway network 1956 COLIN BOOCOCK

Railway construction in Ireland, like that of the earlier canals, was, in civil engineering terms, generally relatively easy. On account of the topography of the island, only a few major bridge structures (viaducts) have been required and even fewer tunnels. In 1859 it was calculated that the average cost of railways in Ireland was only £15,000 per mile compared to £39,000 per mile in England. The lack of commercially viable deposits of minerals, particularly steam coal, meant that the Irish rail network was limited, unlike in Britain, where a plethora of lines resulted from the many manufacturing and marketing opportunities presented by the Industrial Revolution. Nevertheless, there is plenty of railway infrastructure to interest the reader, much of it associated with the main lines of railway, but some with lines long since closed and abandoned.

EARLY IRISH RAILWAYS

Authorisation for Ireland's first railway, the Limerick & Waterford, was obtained in 1826, shortly after the opening of the Stockton & Darlington Railway in the northeast of England, the first railway designed for steam locomotives, though these were confined to goods trains. However, this was too early for a lengthy route to be built in Ireland and it was not until 1834 that Ireland's first railway, the **Dublin & Kingstown (Dun Laoghaire) (D10)** (D&KR), was opened. The decision to build this line was influenced by the opening of the Liverpool & Manchester Railway in 1830.

Charles Blacker Vignoles (1793–1875) was appointed Chief Engineer to the D&KR in 1832 and the contractor selected was William Dargan (1799–1867). Vignoles, born near Enniscorthy, had begun his long and distinguished civil engineering career around 1824 and was to become associated with a number of railway projects in Ireland. He is probably best known internationally as the designer of the suspension bridge spanning the River Dneiper in Kiev. Carlow-born Dargan had worked from 1819 on road construction in England for Thomas Telford and returned to Ireland in 1824 to supervise the construction of the road from the new harbour at Howth to Dublin. He stayed on in Ireland where he in time built up an extensive contracting business, being involved with much of the development of the mainline railway systems on the island. In 1832, Dargan, against stiff opposition, won the contract for building the D&KR. The civil engineering problems faced included the design and construction of sea embankments and protection works and the need to replace a newly built bridge over the River Dodder following a flood. The original two-arch masonry bridge was replaced first by a temporary timber structure, then by an iron bridge by Fairbairn, which subsequently was strengthened in 1883 by the insertion of a central girder. In December 1834, the first regular train was hauled out of Westland Row (now Pearse) Station 'en route' to a temporary station near Dunleary (a fishing village, now part of Dun Laoghaire). The line, which was constructed to a gauge of 4ft 8½in (which

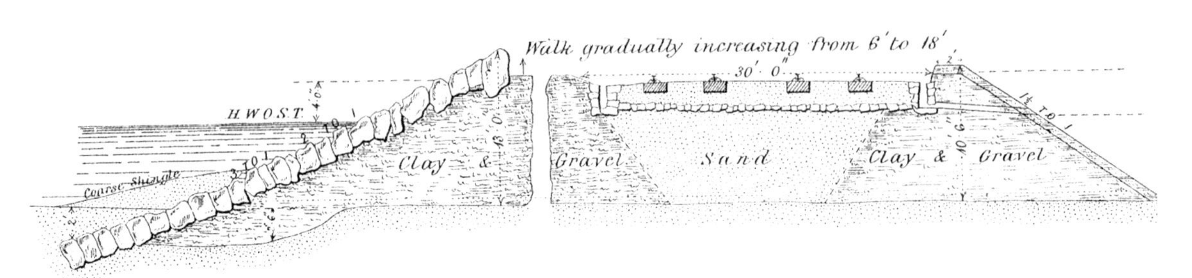

Sea embankment at Booterstown on the Dublin–Kingstown railway

became the standard British gauge), was shortly afterwards extended to a permanent station, designed by the architect John Skipton Mulvany, with a spur to service a new packet (mail boat) wharf at Kingstown Harbour (the mails were transferred from Howth to Kingstown in 1826). From Dublin (Pearse) the line was built as far as Ballsbridge on an embankment enclosed between retaining walls, and then on an embankment that gradually reduces in height as it approaches sea level at Merrion. The line is then carried to Blackrock across an inlet of Dublin Bay on a substantial embankment designed by Vignoles. Between Blackrock and Seapoint, the contractor had to form cuttings through solid rock, resulting in the company having to pay substantial compensation to the local landowners. The landowners also demanded fishing and bathing lodges, a *camera obscura* tower, a small harbour with a boat slip, and an iron-latticed footbridge leading from their properties across the railway to the sea, all the building work being 'executed in the very best style of Italianate architecture'. When the line was extended to Kingstown, another short embankment cut off access to a small fishing harbour and the company had to provide a compensation harbour by building a new jetty on the seaward side of the line.

In the 1840s, Clegg and Samuda demonstrated a method of railway propulsion using air pressure in a tube to propel a piston attached to a rail vehicle. Under the direction of Vignoles, this atmospheric system of propulsion was adopted by the D&KR for their English standard gauge line extension to Dalkey, built by Dargan and opened in 1844. The route, much of it following the line of the tramway used to convey stone for the building of Kingstown Harbour from quarries at Dalkey, involved considerable amounts of rock excavation and substantial retaining walls, especially along the seafront at Kingstown (now Dun Laoghaire). This very successful line was later converted to conventional railway operation when the company was taken over by the Dublin & Wicklow Railway (D&WR), who regauged the line to the Irish standard of 5ft 3in.

THE ROYAL COMMISSION

Charles Blacker Vignoles (1793–1875)

Proposals for further lines were then held in abeyance whilst a Royal Commission was appointed (in 1836) to propose a general system of railways for Ireland. The engineers consulted were Vignoles and Sir John Macneill (1793–1880), Vignoles considered possible routes in the southern half of the island and Macneill those in the northern half. One of their main recommendations was the provision of a rail link between Dublin and Belfast. They further recommended in their final report (1838) the construction of two main lines from Dublin, one to the northwest and the other to the southwest, and that the line from Limerick to Waterford, already approved in 1826, be progressed. The Commission considered that a line to the west was not necessary at the time as the western region was deemed to be served adequately by the Royal and Grand canals.

Macneill was born in Dundalk and gained extensive experience working in Ireland, Scotland and England working on roads, canals and railways, including being assistant to Telford on the London–Shrewsbury road, before returning to Ireland. He built up a substantial practice as a consulting engineer and was engineer to a number of railway companies, one of his largest commissions being the design of the main line between Dublin and Cork for the Great Southern & Western Railway (GS&WR).

LINKING BELFAST WITH DUBLIN (BELFAST TO PORTADOWN AND DUBLIN TO DROGHEDA)

Earlier, the Ulster Railway (UR) from Belfast (constructed initially to a gauge of 6ft 2in) reached Lisburn in 1839. Lurgan was reached by 1841 and Portadown the following year. There were no real problems of a civil engineering nature, apart from contending with some difficult ground conditions near Portadown and the bridge over the River Bann. Meanwhile, the **Dublin & Drogheda Railway (L15)** (D&DR), built to a gauge of 5ft 2in, had opened to Drogheda in 1844. Following the adoption of different gauges by these railway companies, the matter was referred to the Board of Trade who recommended an Irish gauge of 5ft 3in, and this was given legal status by the Railway Regulation (Gauge) Act of 1846.

The first section of the coastal route north from Dublin crosses a number of river estuaries, in particular the Broadmeadow Estuary to the north of Malahide. The engineer, John Macneill, elected to carry the railway across the estuary partly on an earth and rock embankment with an eleven-span viaduct at the southern end. The original timber viaduct consisted of eleven 55ft spans, the abutments and each truss being founded on timber piles. The other viaducts over the river estuaries at Rogerstown, Gormanstown and Laytown were also built originally in timber. By comparison, the substantial viaduct at **Balbriggan (L16)** of eleven 30ft-span arches was built in masonry. The line to Drogheda was opened on 24 May 1844 by the Lord Lieutenant and on the occasion Macneill received a knighthood for his services to the community. The Dublin terminus of the line, Amiens Street (now Connolly) Station, was completed two years later, but was later extended. The station is at a high level and the tracks are carried northwards on a viaduct of seventy-five arches as far as the Royal Canal.

Sir John Macneill (1793–1880)

MASONIC LODGES 212 AND 384, DUNDALK

ENGINEERING OF VIADUCTS

Early on, railway tracks were kept as level as possible and avoided steep gradients because of the limited haulage capacities of steam locomotives at the time. That, of course, meant expensive and labour-intensive engineering works to create deep cuttings, high embankments, bridges and viaducts to carry the line over deep valleys and existing routes. In these early years of railway construction, engineers faced with the need to design bridges and viaducts to carry the rails over physical obstructions, such as river valleys, had the option of using the tried and tested method of masonry arch construction, or to use iron, either cast or wrought or a combination of the two forms. Apart from its basic characteristics of being strong in compression and weak in tension, cast iron contains impurities and is quite brittle, due to its high carbon content. After the Royal Commission of Inquiry of 1850 the use of cast iron in railway bridge construction was largely discontinued and wrought iron became the preferred material. One of the few instances where cast-iron arches were used in a railway viaduct in Ireland survives at **Chetwynd (M18)** southwest of Cork. The viaduct carried the now disused Cork & Bandon railway over the main Bandon road. Wrought iron, on the other hand, is produced by working the iron to drive out the impurities and to reduce the carbon content. In time wrought-iron I-beams and other sections became widely available. The production of wrought iron allowed the construction of large latticed girders and the fabrication of trusses, such as the Pratt truss, of which there are a number of large span examples in Ireland. A truss is made up of many small-section structural members arranged in such a manner that the tension and compressive forces within the members balance each other out across the truss. Trusses normally exhibit a rectangular form in elevation, the arrangement of members determining the type. The general arrangement was to cross-brace these at the bottom members such that traffic passed between the side trusses. The deck carrying traffic may be supported from the bottom of a truss or girder (known as the through type) or across the top of the truss or girder.

Towards the end of the nineteenth century, the railway companies in particular tended to favour the use of solid plate girders in bridges. These could be brought to site in sections and there riveted or bolted together. Many have been removed with the closure of a large number of railway lines but some remain. Later plate girders were made of steel, especially after it became possible to roll sheets and sections such as angles. Plate girder bridges, although of poor aesthetic quality, tend to be cheaper, especially when replacing earlier trusses on existing abutments. They also have a robust appearance, which gave a sense of security to rail passengers.

LINKING DUBLIN WITH BELFAST (DROGHEDA TO PORTADOWN)

The Dublin & Belfast Junction Railway (D&BJR) was formed in 1844 to promote and provide the necessary commercially important link between Drogheda and Portadown passing through Dundalk, so meeting the original ambition of a main rail route between Dublin and Belfast. Construction was extremely slow due to geographic and topographical

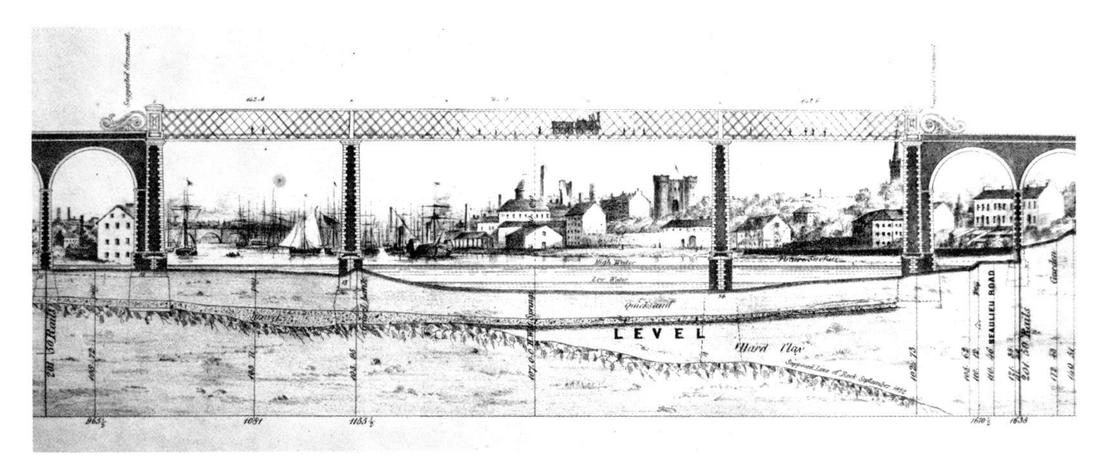

Part of Boyne Viaduct contract drawing SCHOOL OF ENGINEERING ARCHIVE, TRINITY COLLEGE DUBLIN

Egyptian Arch under Belfast–Dublin railway, near Newry, County Down

Carlow Station

constraints, not least of which were the high-level crossing of the Boyne Estuary at Drogheda, and a high viaduct near Newry, some deep cuttings in granite, and substantial embankments. A through connection was not finally established until 1855 with the completion of the **Boyne Viaduct (L17)**. This imposing viaduct and bridge over the estuary of the River Boyne at Drogheda, about 32 miles north of Dublin, was constructed in its original form between 1851 and 1855 as the final link in the rail route between Dublin and Belfast. The Admiralty, as they had done at the Menai Straits between Anglesey and mainland Wales, insisted on a minimum headroom of 90ft over a minimum clear waterway of 250ft. To meet these conditions, the consulting engineer, Sir John Macneill, chose to use wrought-iron double-latticed girders for the main spans, the detailed design of these being left to James Barton, Chief Engineer to the railway company. The principles of multi-lattice construction in wrought iron were first applied on a large scale on the Boyne Viaduct contract and led to a significant increase in the knowledge of the structural properties of wrought iron and encouraged more exact computation of the stresses in structural components. Increasing axle loads led in 1932 to the replacement of the latticed girders with simply supported mild steel girders, the central span being given a curved top chord, more to improve its appearance than for any structural reasons. This is how the viaduct looks today.

Further north, the **Craigmore Viaduct (U22)** near Newry curves slightly in plan at its northern end and has eighteen semicircular masonry arches, each of about 60ft span. The height of the permanent way above the lowest part of the Bessbrook River valley below is 137ft, making it the tallest railway bridge in Ireland. It was built of local Newry grano-diorite, designed by Macneill, and built by Dargan. The present station for Newry is located to the west of the town near the southern end of the viaduct. A feature of this part of the line is the considerable amount of rock excavation that was required, and the erection of a number of skewed masonry arches of remarkably thin section.

DESTINATION CORK

As engineer to the Great Southern & Western Railway (GS&WR), Macneill served a company that grew to be the largest in Ireland, with over 1,100 miles of track, 240 miles of which was double track. In partnership with William McCormick, Dargan first completed the line from Dublin to Cherryville Junction and Carlow, his home town, and then extended the main line as far as Maryborough (Portlaoise). With the aim of effecting the connection with Cork, construction reached Thurles by 1846, but the individual contractors then faced serious financial and other problems, occasioned largely by the effects of the Great Famine. At this stage, Dargan, who had by then built up an enviable reputation for quality work, was awarded a single contract to complete the construction of the line to Cork. Having crossed the River Blackwater near Mallow, the line enters north

County Cork. Here, the terrain became more difficult and required much heavy engineering of cuttings and embankments. Dargan constructed a magnificent ten-arch cut-stone viaduct over the River Blackwater near Mallow, but sadly it was blown up by anti-Treaty forces during the Civil War in 1922, and replaced the following year using steel girders. Near Rathpeacon, the existing Monard and Kilnap masonry arch viaducts brought the main line in 1849 to within a mile of the River Lee at Penrose Quay. From a temporary terminus at Blackpool, it was to be another six years before Dargan was able to complete a tunnel to bring the railway to a terminus near the city. Later, the present terminus at Lower Glanmire Road was constructed closer to the entrance to the tunnel. The tunnel, to Macneill's specifications, is 4,065ft long on a downgrade of 1 in 60, is 24ft high and generally 28ft wide and is the longest operational rail tunnel in Ireland. Its width increases at the southern end to allow the twin tracks to diverge as they approach Kent Station. The **Rathmore Rail Tunnel (M16)**, driven through a sandstone ridge, has four ventilation shafts, is lined throughout with a three-ring brick arch springing off the rough masonry walls. Ireland's longest rail tunnel was, however, far to the north, at Lissummon on the Newry & Armagh Railway. The line was quite hilly and required two tunnels, that at Lissummon being just short of 1 mile, the other at Loughgilly being 1,095ft long. Although the branch was closed in 1955, the longer tunnel still exists, although now largely derelict.

William Dargan (1799–1867)
IRISH RAILWAY RECORD SOCIETY

RAILS ACROSS THE BOGS

Whilst rails were being laid southwards from Dublin, the Midland Great Western Railway (MGWR) was formed in 1845 with ambitions to reach Galway and other areas to the west and northwest of the country. The MGWR bought out the Royal Canal Company, appointed George Willoughby Hemans (1814–85) as chief engineer, and Dargan obtained the contract to build the main line from Dublin following the general line of the canal as far as Mullingar. Hemans, born near St Asaph in north Wales, had begun his career in Macneill's office in London before, in 1840, being appointed resident engineer for the construction of the D&DR (opened 1844). Between Enfield and Mullingar, Hemans elected to cut straight across the bog rather than follow the sinuous curves of the canal. He experienced considerable difficulty in building the railway through the bog and is usually credited with developing a viable method of construction that overcame the problems. Such bogs could reach depths of up to 60ft in places. Canals and roads had previously been constructed in this type of terrain and the experience of the early engineers pointed to a need for significant pre-construction drainage of the area of the proposed route. A system of drains all along and across the intended line was installed and, when this drainage system was working well, the level of the bog formation gradually subsided. The next stage in the process was to lay two courses of heather sods to a width of 30ft on which it was intended to seat the rails, all holes and inequalities in the surface of the bog having been first levelled with peat. The permanent way on this formation consisted of half baulks of red pine, about 25ft long, on top of which were 12in by 6in timbers laid transversely at intervals of 3ft and over these were laid the longitudinal timbers also measuring 12in by 6in. The flat-bottomed ('Vignoles') rails were bolted to the longitudinal timbers, the bolts passing through the transverse bearers and secured on the underside. A layer of dry, dense peat, either on its own or mixed with fine gravel, was then placed, which provided material for packing the track to the required line and level. Modern problems of maintaining good railway running conditions on the tracks in bogs relate to differential settlement, vibration of the track under trains and formation failure. The main solution to the vibration and differential settlement problems is to increase the weight of the formation without causing formation failure. The first train reached Mullingar, 36 miles from Dublin, in October 1848, the 76-mile stretch between Mullingar and Galway being completed by August 1851. Major engineering features of the line include the crossing of the River Shannon at Athlone, the River Suck near Ballinasloe, and Lough Atalia on the approaches to Galway city. The **Athlone Rail Bridge (L18)** was completed by Fox Henderson & Company in 1851, and is formed from twin 176ft-span wrought-iron latticed bowstring girders and a central cantilevered opening span of 117ft (permanently closed in 1972) in addition to two shorter side spans (renewed in steel in 1927). All the girders are carried on piers formed of cross-braced cast-iron cylinders, compressed air being used when sinking the cylinders into the riverbed, the first time in Ireland that this technique of excluding water from the workings was employed.

RAILWAY STATION ARCHITECTURE

The four great lines of railway in Ireland developed separately and each had a major terminus in Dublin. Each railway company built an imposing, solid and impressive terminus building in conservative classical styles with the objective of impressing the investors and reassuring the public at a time when the railways were new and somewhat alarming. Each had a substantial train shed with a large-span roof covering the platforms, the roofs being generally of wrought iron supported on cast-iron columns.

The architect John Skipton Mulvany (1813–70) designed the main stations on the D&KR (Blackrock and Kingstown (Dun Laoghaire)), but the first station on the line (**Westland Row (D11)**, now Pearse Station), was an unassuming design produced by Vignoles. The present station roof design was adapted from that of Lime Street Station in Liverpool when the roof was rebuilt in 1889. When opened in 1846, Amiens Street (Connolly) Station, the terminus of the D&DR, was the most impressive railway building in the country. Designed by William Deane Butler (*c.* 1794–1857) in the Italianate style and built of Wicklow granite, its most striking feature is the central tower. The northern terminus of the UR at Great Victoria Street in Belfast was an imposing building, with a Doric colonnaded portico, but perhaps the finest of the Belfast termini stations was York Road. Originally a modest building, it was extended in the 1890s to the plans of the engineer to the Great Northern Railway of Ireland (GNRI), Berkeley Deane Wise (1853–1907). Wise hailed from New Ross in County Wexford and was engineer to the Belfast & County Down Railway (B&CDR) and a number of other railway companies during his career. **Kingsbridge (D12)** (now Heuston) Station in Dublin is regarded as one of the city's finest buildings. Designed by Sancton Wood (1815–86), an experienced English railway architect, the building, in Corinthian style, consists of a central front flanked by wings each surmounted by a tower. The train shed, designed by Macneill and opened in 1848, originally covered four platforms, the roof being supported on cast-iron pierced spandrel arches spanning between rows of cast-iron columns, the ironwork being supplied by the Dublin firm of J. & R. Mallet. The riverside wing was added in 1913 and in recent years, further platforms were added, the roof members replaced in steel, and modifications made to the public concourse. Sancton Wood is also credited with the main administration buildings at the Inchicore works a few miles west of Heuston Station and the stations between Monasterevin and Limerick. The **Inchicore Railway Works (L18)** is the sole surviving example of an independent railway works in Ireland, and one of the earliest in the world. Macneill also designed the stations between Dublin and Carlow, which are in the Elizabethan style.

The MGWR employed Mulvany, the architect of the D&KR, to design most of its principal stations between Dublin and Galway. The two termini and the Great Southern Hotel at Galway in particular are fine examples of Mulvany's work. Broadstone, the original Dublin terminus of the MGWR built near the terminus of the canal, is built of granite, combining Grecian and Egyptian styles of architecture with an elaborate Doric style colonnade on its east side added later to a design by the architect George Wilkinson (*c.* 1814–90). The roof of the adjoining train shed was supplied by ironmaster Richard Turner (better known for his Victorian glasshouses) and was replaced around 1970.

WATERFORD TO LIMERICK

Dargan was also the contractor (possibly with his brother James) for much of the line between Waterford and Limerick (W&LR), which opened between Limerick and Tipperary in May 1848. The section of the line between Tipperary and Waterford (reached in 1854), included the crossing of the River Suir at Cahir in County Tipperary. For the **Suir Viaduct (M23)** (completed 1855), the engineer to the W&LR, William Le Fanu, specified wrought-iron box girders spanning between masonry piers and abutments. The girders were supplied by William Fairbairn of Manchester, the then holder of a patent for such girders, and represent the earliest, largely unaltered, example in Britain or Ireland. The track (permanent way) is carried on transverse beams, originally spanning between the box girders, but now, following damage resulting from an accident, suspended by steel frames from the tops of the main girders.

SOUTHWARDS FROM DUBLIN

Southwards from Dublin, the D&WR mainline commenced at Harcourt Road (the line was later extended to a fine terminus at Harcourt Street designed by George Wilkinson), whilst the coastal route from the former D&KR at Dalkey joined it at Shanganagh Junction prior to crossing the Dargle River to reach Bray in 1853. The so-called Harcourt Street Line crossed the River Dodder and the Shanganagh River valleys on multi-span masonry arch viaducts, the first of which has been reused recently to carry the Green Line of the LUAS light-rail service. It has been proposed to reuse the other viaduct in a similar way for a future extension of the LUAS to Bray. The consulting engineer to the D&WR, appointed in 1845, was none other than the famous English engineer, Isambard Kingdom Brunel (1806–59). The tunnel between Dalkey and Killiney has a typical Brunel cross-section with heavy portals. It was driven through solid granite and is lined throughout. The embankments and retaining walls further to the south along the coast also presented a challenge. However, construction of the line onwards from Bray towards Wicklow provided even more of a challenge. The first 5 miles along the sea cliffs of **Bray Head (L19)** required the driving of three tunnels and the erection of four timber trestle viaducts, all designed by Brunel and completed by Dargan. The Precambrian rocks of the headland are generally unstable and there have been a number of deviations of the alignment and two of Brunel's tunnels have been relined and the other bypassed. A final major deviation became necessary and a tunnel, 3,252ft in length, was completed in 1917. Further south, between Greystones and Wicklow, the line was built along the edge of the foreshore and has always been subject to considerable coastal erosion, posing problems for the civil engineering staff. Under the guidance of the Dublin, Wicklow & Wexford Railway (DW&WR) company, construction of the route continued southwards towards Wexford, but progress was slow due to the

Building a masonry arch rail viaduct c.1860

difficult terrain in the Rathdrum and Avoca areas of County Wicklow, and Enniscorthy in County Wexford was not reached until late in 1863. The two Rathdrum viaducts span the Avonmore River, and consisting of a number of 44ft semicircular arches, were completed in 1861. The early photograph shown here offers an indication of how these masonry viaducts were built. The DW&WR, having become the Dublin & South Eastern Railway (D&SER) in 1907, continued the line from Enniscorthy following the southwestern bank of the River Slaney, and finally reached Wexford in 1872.

AN EXPANDING NETWORK

Meanwhile, the MGWR was extending its network to serve centres lying to the northwest of Dublin, such as Navan, Longford, Sligo, Ballina, and Westport. River systems were crossed by multi-span masonry arch viaducts, that spanning the River Boyne at Navan being a typical example, or by iron truss bridges of varying design. The iron bridges on the main routes have needed to be replaced in recent years to allow for the upgrading of the lines to accommodate higher loadings and speeds. However, the major viaducts have tended to survive and have been refurbished and strengthened, an example being the viaduct crossing the valley of the River Nore at **Thomastown (L22)** on the line from Kilkenny to Waterford. The viaduct is a solid Victorian bowstring wrought-iron girder bridge of 212ft span completed in 1867.

Wellingtonbridge Station footbridge, County Wexford

To the southeast, an Act of 1894 empowered the construction of the Rosslare Strand–Waterford line and the absorption of the Waterford, Dungarvan & Lismore (D&LR) and the Fermoy & Lismore (F&LR) railways (completed in 1878). The GS&WR, having already constructed the Mallow–Fermoy branch in 1860, a new route from Rosslare to Cork via Waterford and Mallow was thereby provided. Between 1904 and 1906, Sir Robert MacAlpine constructed the 36-mile line between Rosslare Harbour and Waterford, including a major viaduct across the River Barrow west of Campile and another near Wellingtonbridge. The **Barrow Viaduct (L21)** near the confluence of the Barrow and Suir river systems was designed by Sir Benjamin Baker and is the longest rail bridge in Ireland spanning entirely across water. The bridge consists of thirteen mild-steel Pratt-type latticed girder spans and an opening span, all supported on cross-braced cast-iron cylindrical piers. The contractors for the line made extensive use of concrete and brick in the construction of the various bridges and viaducts, that near Wellingtonbridge being the most substantial.

On the line between Waterford and Dungarvan, the only structures of note are the curving masonry River Mahon Viaduct at Kilmacthomas, with its eastern approach protected by a high retaining wall, and the Ballyvoile Viaduct over the River Dalligan in the southwest of County Waterford. The line and its structures were completed in 1878 and formed part of the Waterford, Dungarvan & Lismore Railway (WD&LR), later absorbed by the GSWR and operated by Irish Rail as freight-only until the late 1980s. A section of this line between Kilmeaden and Waterford is now operated by a heritage group. The original masonry arches of the Ballyvoile Viaduct were destroyed in the Civil War and rebuilt in 1923. The present structure consists of four spans of twin 78ft-long Pratt-type trusses supporting the deck and carried on 'mass concrete' piers. Once over the Colligan River at Dungarvan, the next river system to be crossed was the so-called 'Munster' Blackwater at Cappoquin. Here the railway was carried across the river on a viaduct

Kilcummer Rail Viaduct near Fermoy, County Cork *c.* 1860

consisting of a number of plate girder spans approached on each side by a series of masonry arches.

Before reaching Fermoy, the line again crosses the Blackwater, this time on a viaduct consisting of a main river span and five side spans. The **Carrickabrack Viaduct (M22)** was renewed during the period 1903–06. The central 157ft span of twin latticed girders is approached across the northern flood plain of the river by five 49ft spans of reversed parabolic steel plate girders with rolled-I cross girders to carry the single track. On the original GS&WR branch line from the main line at Mallow to Fermoy, the line crosses the valley of the Awbeg River, a tributary of the Blackwater, on a high viaduct of ten spans (total length: 465ft). Again, double latticed wrought-iron girders were used to span between the tapered masonry piers, the girders being spaced 5ft 3in apart (the rail gauge). The girders are probably early-twentieth-century replacements for the originals. Unlike many others, the Kilcummer Viaduct may be viewed with ease from the grounds of nearby Bridgetown Priory.

DOWN TO VALENTIA

In the 1850s, the GS&WR had opened up the southwest county of Kerry with their route from Mallow to Killarney and Tralee and, by 1885, had built a branch from Farranfore Junction to Killorglin, where a fine example of a bowstring truss viaduct carried the railway across the River Laune. The **Laune Viaduct (M19)** is of five spans, two masonry approach spans, one to each side, and three river spans, each of 95ft. The line was closed in 1960 and in late 1993 timber decking was added and the viaduct reopened as a public footpath forming part of a circular riverside walk. The government of the day was considering the possible use of Valentia as a point of departure for the Atlantic mails and financed, by means of grants and Baronial guarantees, the continuation of the railway at standard gauge along the southern shore of Dingle Bay. Opened to Valentia in 1893, this 16-mile section included some of the steepest gradients and most spectacular civil engineering works on any Irish railway. The permanent way was often formed on a narrow shelf cut into the hillside with the road and sea far below. There were a number of tunnels and heavy earthworks, including large retaining walls, and a curving viaduct at **Gleensk (M20)**. This viaduct was built on a 650ft-radius curve and has eleven 58ft spans formed of steel girders.

NARROW GAUGE RAILWAYS

In the most sparsely populated areas of Ireland, it was not possible to justify the cost of construction of standard gauge lines and this led to the passing of a number of Acts in the late nineteenth and early twentieth centuries authorising short lengths of 3ft-gauge railways, which together became the most extensive narrow gauge railway system in either Britain or Ireland. In all, some 550 miles of narrow gauge lines were built, mainly in counties Galway, Kerry, Mayo and Donegal, to serve remote and thinly populated districts where, as commercial ventures, they could not have been justified but later proved to be of great benefit to the country. Little remains today of the railways, and it is hard to imagine how they once served well the scattered population in areas where the roads were generally poor. Apart from low revenues, a collapse of fishing during the First World War and competition from road transport eventually sounded the death knell of the narrow gauge lines, especially in County Donegal. Due to the benign nature of the topography, the civil engineering of the narrow gauge lines was not difficult, the viaduct at Owencarrow proving to be one of the few challenges to be met, but today only the piers remain.

CORK LINES

Like Dublin and Belfast, Cork was served at one time by a plethora of lines operated by several different companies. Some were able to approach the city by way of the valley of the River Lee, others required tunnels to overcome the intervening high ground to the north and southwest. Mention has already been made of the approach tunnel from the north. What later became the Cork, Bandon & South Coast Railway (CB&SCR) needed a tunnel at Gogginshill to access their terminus at Albert Quay, and earlier had driven Ireland's first rail tunnel at Kilpatrick between the Brinny and Bandon rivers. The heavy engineering on this line also included tunnels at Ballinhassig and Innishannon and a viaduct of two 105ft spans over the Bandon River, originally formed of American-style timber trusses, but replaced with wrought-iron latticed trusses in 1862.

CONCRETE VIADUCTS

Although a number of railway bridges were rebuilt in the early twentieth century using plate girders, iron was soon displaced by concrete (apart from one or two larger bridges). Concrete has the same disadvantage as cast iron – it is strong in compression but weak in tension. Thus, initially, its use was confined to foundation work and as an infilling or 'hearting' to bridge piers and arches. Thus it was used as the 'hearting' in the piers of the **Ballydehob Viaduct (M21)** in west County Cork in 1886, in the Taylorstown Viaduct

near Wellingtonbridge in County Wexford in 1906, as well as on the construction of the **Tassagh Viaduct (U15)** and some bridges of the Armagh & Castleblaney Railway through Keady in County Armagh. To the north of Belfast, the construction of the **Greenisland (B14)** loop line in the early 1930s involved the greatest concentration of concrete structures in Ireland. The works included four under- and four overbridges, all of which are different, and two large viaducts, all constructed in concrete. The largest structure, the main line viaduct, has three main arches of 83ft clear span and nine approach

Tassagh Viaduct, County Armagh

Installation of a 70-tonne match-cast concrete unit on the Cross-Harbour rail viaduct, Belfast
ESLER CRAWFORD PHOTOGRAPHY

arches, each of 35ft span. The structures represent an important stage in the development of the use of reinforced concrete for railway work. In the 1990s, the isolated line to Larne was linked with Central Station by the completion of major road and rail viaducts over the River Lagan (**B2**). The rail structure, named for William Dargan, with a length of 1,490m, is the longest rail bridge structure in Ireland, being over twice the length of the Barrow Viaduct near Waterford. The Dargan Viaduct has 47 spans, the deck consisting of 360 matched precast segments built off the piers as balanced cantilevers, the units being post-tensioned after placement.

CONCLUSION

The early railway engineers must be given credit for the amount of time and effort they expended and the vision they had of the development of future transport needs. Of necessity, the only means they had in getting around to their various sites was on horseback or in horse-drawn vehicles. Yet they carried out extensive surveys and investigated alternative alignments in their feasibility studies, which involved them in long hours of travel and absence from home and family. The engineers and navvies completed the many cuttings and embankments on these early railways at a time when mechanical methods of transportation and excavation were unavailable. They did, however, have the benefit of the experience of the canal engineers, who, half a century earlier carried out the immense feat of building the Grand and Royal canals across similar very difficult terrain. The upgrading of much of the railway infrastructure is continuing. A major consequence has been the replacement of many of the stone arch overbridges with structures having a more rectilinear profile to improve clearance for high-speed train movements and the refurbishment, strengthening or total replacement of bridge structures and viaducts at some river crossings.

WATER AND DRAINAGE

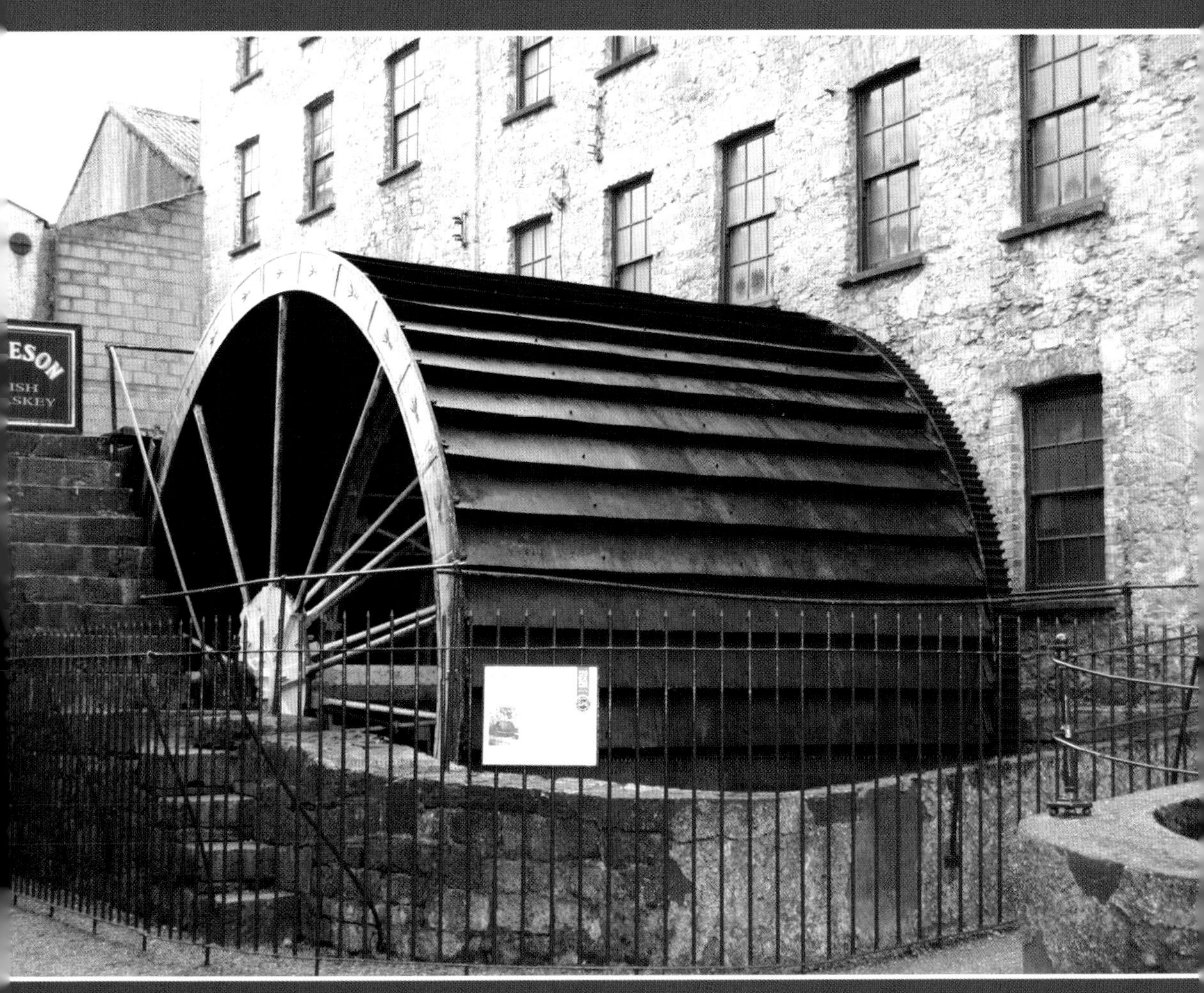

Water wheel at Midleton distillery, County Cork COLIN RYNNE

Civil engineers have been, and still are, intimately involved in the provision of fresh drinking (potable) water, the removal and treatment of sewage and wastewater, arterial drainage and the harnessing of Ireland's rivers to generate hydroelectric power. Although these activities will here be dealt with separately, they are not independent. The disposal of sewage must be accomplished without polluting freshwater sources, and the construction of hydroelectric dams on rivers has often provided opportunities to improve the freshwater supplies to urban areas. In a similar manner, river improvement schemes were often undertaken with the dual purpose of improving both navigation and drainage, and for many years the Grand and Royal canals provided Dublin with drinking water. Even waterpower and canals can be related, for example the many leats (trenches or ditches that convey water to a mill wheel) at the Ballincollig gunpowder mills near Cork doubled as an internal canal system for the distribution of materials.

PROVISION OF POTABLE WATER

Potable water is one of the basic requirements for human habitation, agriculture and industry. The principal components of a modern water supply system comprise a reliable source of water, a system of treatment that ensures the water is safe to drink, elevated storage facilities to provide a public supply at sufficiently high pressure, and a distribution network of pipes.

The history of hydraulic engineering stretches back to ancient times and some of the ancient aqueducts of the Romans and the qanats (gently sloping irrigation tunnels) of Iran are still in use. In Ireland the prehistoric constructions at Newgrange, Knowth and Dowth, which date from approximately 3000 BC, show that the early Irish had the social organisation and technical ability to construct significant structures requiring the quarrying and transportation of large stone slabs over many miles. However, there is no evidence of significant water engineering in Ireland prior to the eighteenth century. This is due principally to Ireland's topography and climate. The majority of Ireland's mountains are located around the coast and the centre of the country is quite flat, the country often being described as saucer shaped. The annual rainfall in Ireland varies between 750 and 1,000mm on the eastern side of the country and between 1,000 and 1,250mm in the west, with mountainous regions receiving in excess of 2,000mm per annum. Ireland's rainfall is relatively evenly spread throughout the year and evaporation rates are low. As a result the rivers that drain Ireland's central plain – the centre of the 'saucer' – are slow moving and flow all year round. Thus the earliest settlements, which were typically situated on rivers or lakes, enjoyed a steady year-round water supply and there was no necessity to build aqueducts, or to provide means of storing water during dry seasons.

Much of rural Ireland's drinking water is still supplied from wells, either individually owned, or supplying a group scheme. Initially, towns were also supplied by wells or directly from the rivers on which they were founded. The need to develop engineered water supplies arose only in the bigger cities and towns as their populations grew.

The earliest Irish public water supply dates back to thirteenth-century Dublin when a feed was taken from the River Dodder at a weir at **Balrothery (D23)**. Tapping a river upstream of a town was prudent, given that the rivers and streams drained the towns and washed animal and human waste downstream. This pattern of river use persists to the present with water being drawn from rivers upstream of towns and outflow from sewage treatment plants being released into rivers downstream.

Medieval public water cistern in Dublin DUBLIN CITY ARCHIVES

Until the mid-nineteenth century, most water supplies in Ireland and elsewhere were untreated. This is not to imply that water quality was never considered. The Roman author Vitruvius gives detailed and sensible advice on identifying and testing freshwater sources while Frontinus, who was appointed *curator aquarum* of the city of Rome in AD 97, distinguished between the water quality of Rome's different water sources and took steps to improve the quality and reliability of the water system. Water filtration was sometimes used to remove particulates, but the importance of water as a disease vector was not recognised until the latter part of the nineteenth century. In 1854 Dr John Snow identified a pump in Broad Street in the city of Westminster in London as the source of a local cholera outbreak. He established the link between the water source and the outbreak by removing the handle from the public pump, whereupon the outbreak ceased. However, his findings were not accepted until decades later when the research of Louis Pasteur, Robert Koch and others confirmed 'germ theory'.

Irish cities were among the first to introduce sand filters to remove particulate matter. Although it was not known at the time, such filters also help to remove biological contaminants. With the acceptance that water was a vector for disease the need to remove or kill water-borne pathogens became clear. Chlorination was first used in Reading in England in 1904 and was quickly adopted in Ireland. In the 1920s, physico-chemical processes for treating turbid and coloured water were introduced. These processes involve the addition of chemicals that cause very small particles to coagulate into larger flocs, which are easier to remove. Irish water is now also fluoriated to help reduce tooth decay.

URBAN WATER SUPPLIES

DUBLIN

Dublin's engineered water supply dates back to the thirteenth century when part of the flow from the River Dodder was diverted into a canal, 2 miles long, that fed into the River Poddle. Up until that time, the Poddle had been the city's traditional source of drinking water. The flow from the Dodder was taken upstream from the Balrothery Weir under an agreement with the Priory of St Thomas, whose monks had constructed the weir. The augmented flow from the Poddle was diverted into a cistern near St James's Gate in the city. This cistern was replaced by the 'City Basin' in 1721. A new supply for the north city

was established in 1741 by extracting water from the River Liffey above the weir at Islandbridge (downstream of the weir the Liffey is tidal). This supply involved pumping water from the river.

The construction of the Grand Canal began in the late 1750s. The original, and ultimate, purpose of the canal was to link Dublin with the River Shannon. However, Dublin Corporation's Pipe-water Committee realised the canal's potential as a water supply. As a result, the Corporation put up money towards the completion of the section of the canal from Dublin to the River Morrell and took over administration of the works. Eventually, the works were taken over by the Grand Canal Company, which was supported by the Corporation, and in 1777, two years before the canal was opened to traffic between Dublin and Sallins, the canal began supplying water to the City Basin. Another basin was subsequently added to the Grand Canal system at Portobello.

The canal-fed system on the south side and the Islandbridge supply on the north side were both low-pressure systems. A new storage basin, which was connected to the Royal Canal, was added at Blessington Street in 1814, but by the 1850s the city's water supply was becoming inadequate.

The Vartry system, which takes water from the **Lough Vartry (L27)** impounding dam and reservoir near Roundwood in the foothills of the Wicklow Mountains, was completed in 1867. Water from the reservoir at Roundwood is transported via an aqueduct, 33 miles long, to a service reservoir at **Stillorgan (D24)**; 2½ miles of the aqueduct comprises a tunnel cut with considerable difficulty through solid rock. The Vartry supply also serves Wicklow, Bray, Dun Laoghaire and other areas of south County Dublin. A severe drought in 1883 indicated the need for increased storage at Roundwood, and this was provided by the completion of a second dam and reservoir in 1922. The Vartry scheme supplies water at high pressure.

With the completion of the Vartry scheme in 1867, Dublin Corporation stopped taking water from the Grand Canal. Rathmines, which was a

Eduction tower, Lough Vartry, Roundwood, County Wicklow

Vartry water treatment works near Roundwood, County Wicklow

separate township until 1930, had been taking a supply from the Grand Canal near Clondalkin. However, by the 1880s this low-pressure system was inadequate and a reservoir was constructed in the Dodder valley at Bohernabreena with sand filters at Ballyboden.

By the 1930s, Dublin's demand for water was once again exceeding capacity and an additional supply became necessary. This time the city's water needs were provided as a by-product of another important piece of engineering infrastructure. The dam of the ESB's hydroelectric scheme at Pollaphuca created the **Blessington (L24)** artificial lake and agreement was reached to abstract water to supply Dublin and parts of Kildare. The supply is transported to the city by an aqueduct to Rathcoole from where it is piped to reservoirs at Saggart, Belgard and Cookstown. A second main was added in 1953 and an aqueduct, 25 miles loing, to Fairview was completed in 1994. Water from the Blessington reservoir is treated at the Ballymore Eustace water treatment facility adjacent to the lake.

In 1965 north County Dublin was also provided with an additional source following the construction by the ESB of a hydroelectric dam on the River Liffey at Leixlip.

BELFAST

Belfast is a very new city compared to Ireland's older Viking towns. In 1650 it had a population of only 1,000 people, but between 1800 and 1900, when Belfast underwent rapid industrialisation, its population increased from 20,000 to 350,000, equalling Dublin in size. Before 1795, the water supplies were privately owned. After this date, the Belfast

Ben Crom Dam, County Down

Charitable Society took responsibility for providing the city with water before handing over to the Belfast Water Commissioners, constituted under the Belfast Water Act of 1840. The Belfast Water Commissioners developed a new supply with reservoirs (**B15**) designed by L.L. Macassey in the hills to the north of Belfast Lough. Between 1860 and 1890 seven additional storage reservoirs were built so that by 1900 the Belfast system was supplying 455,000 people.

The increase in demand necessitated the extraction of water from the Kilkeel and Annalong valleys in the Mourne mountains. The water from these rivers was diverted into an aqueduct, 35 miles in length, 6⅔ miles of which is in tunnel, 12⅔ miles in cast-iron siphon pipes and the rest 'cut and cover'. The aqueduct was completed to the Knockbreckan reservoir near Belfast in 1901, but the Silent Valley reservoir on the Kilkeel River was not completed until 1933. In 1957 an additional dam was constructed further up the valley to form the Ben Crom reservoir. Also, in the 1950s, the 2¼-mile-long Slieve Bignian tunnel was driven from the Annalong River to the Silent Valley in order to direct additional water to the lower reservoir.

In 1960 new plans were drawn up to take water from Lough Neagh and these works were complete by 1972. Since then, in 1979, as a short-term measure, boreholes were developed to exploit groundwater in the Bunter sandstone in the Lagan Valley. In addition, following the drought of 1973, water from the Lough Island Reavy reservoir, which was originally constructed in 1839 to facilitate mill owners on the River Bann, was connected to the Silent Valley Aqueduct.

Steam-driven pumps by Combe Barbour, Belfast, at Lee Road Waterworks, Cork

CORK

In 1768 the Cork Pipe Water Company installed a water wheel and pumps at a weir on the River Lee just west of the city. In 1774 a second larger reservoir was constructed beside the first at a height of 65ft above river level. By 1845 this supply had become inadequate, both in terms of capacity and pressure. The Cork Corporation bought the company and installed a new steam pump and constructed new reservoirs at 196ft Ordnance Datum (OD) and 387ft OD. Cork's timber pipes were replaced between 1867 and 1870 and water was supplied to 125 fountains to serve houses that lacked plumbing. At about the same time (1863–69), additional steam pumps were installed. Steam continued to be used as the power source until replaced in 1939 by electricity, stand-by steam plant having been installed between 1905 and 1907. Between 1888 and 1916 five water turbines were installed which drove reciprocating pumps. In 1980 these pumps were still delivering 20 per cent of Cork's water supply.

Water treatment in Cork began in 1879 with the use of an infiltration gallery. In 1956 new upward-flow clarifiers, new filters and a new pumping station were provided. The plant was upgraded in 1975 and a new water tower constructed.

In 1982 a new supply was developed to serve Douglas, Mahon, Little Island, Great Island and Ringaskiddy, areas that have seen significant industrial development, particularly in the pharmaceutical industries. This supply was provided as a by-product of the ESB's hydroelectric dam on the River Lee at **Inniscarra (M30)**.

LIMERICK

In 1834 the London Waterworks Company began constructing a water supply for Limerick. Prior to this there was no organised supply and water was taken from the River Shannon and from wells scattered throughout the city. This system had an intake from the Shannon at Rhebogue. Water was taken from the river using two 20hp steam pumps and discharged into a settling tank before being run through a rough sand filter. After approximately twenty-five years of population increase and wastage from the pipe network, this system became inadequate and led to water being rationed by supplying water to the city on alternate days.

In 1891 Limerick Corporation, who had purchased the rights of the London Waterworks Company in 1885, constructed a new supply at Clareville near the falls of Doonass. This system used the head available in the River Shannon (between 9ft and 10ft at this point), to drive four Hercules turbines operating four double-acting pumps.

The Shannon hydroelectric scheme of 1929 reduced the flow at Clareville significantly, which made the waterworks inoperable, but a new supply was piped to Clareville from the head-race canal upstream of the works. Further improvements and treatments were added subsequently.

Many of Ireland's other large towns have engineered systems dating back to the second half of the nineteenth century. Occasionally, public fountains, such as that in the County Clare village of Ballyvaughan, survive from these early works. In the last thirty years the numerous local pumps that dotted Ireland's roads have all but disappeared as villages have been connected to regional water supply schemes.

WATER TOWERS

Water supply systems are largely hidden from public view. Their pipes are buried and storage dams and treatment facilities are usually located in remote areas. Water towers are an exception. Modern potable water is supplied at high pressure. Reservoirs are usually sited in upland valleys to provide the necessary head, but in flat terrain structures must be provided to elevate the reservoir. Because water towers are located in flat regions they tend to be highly visible.

The appearance of a water tower usually reflects its function and the materials and techniques used in its construction. The majority of Ireland's water towers are constructed in reinforced concrete but a wide variety of forms have been used, from slender 'wine glass' towers constructed using specialised formwork, to early towers constructed using the Hennebique system of reinforcement. Water towers that originally supplied water for steam trains can still be seen at some of Ireland's railway stations.

WASTEWATER

In rural areas one-off houses typically discharge sewage into septic tanks. Up until the middle of the twentieth century it was not unusual for country schools and houses to use outhouses or dry toilets. In the urban setting, the public provision of wastewater works (sewerage) did not begin to develop before the eighteenth century and was not adequately developed until the twentieth century. In the late eighteenth century the majority of residents in Irish cities used chamber pots, commodes and slop buckets that were emptied into covered cess pools to the front of the house or dumped directly into the street. In heavy rains the contents of these pools were washed into the street drains, which typically comprised open trenches. Early drainage works involved improving the open street drains. First, masonry sides were added, later these drains were covered with brick arches and ultimately stone or brick inverts were added.

During the latter stages of the eighteenth and early years of the nineteenth centuries, Dublin's drains were improved on a street-by-street basis. Commissioners with responsibility for drainage identified streets that required improved drains and then imposed a charge on the occupants to cover the cost of implementing the improvements. These schemes were

Facing page: Water tower at Oranbeg, County Galway

Ringsend wastewater treatment works DUBLIN CITY COUNCIL

not always supported by the occupants and some streets refused to cooperate with the commissioners. The drainage that was implemented tended to be piecemeal, with different drain sections being used and with scant regard to invert levels. Sometimes even the direction of the fall was incorrect.

These drains emptied into the nearest watercourse and hence into the River Liffey. They were designed for rainwater run-off, but also transported the foul matter, including human waste, that was dumped into the streets and overflowed from cess pits. During the nineteenth century the use of water closets became more prevalent. These also discharged into the drainage system so that Dublin's system was a combined system that carried rainwater run-off and sewage in the same conduit. The effect of these drainage works was to wash sewage directly into Dublin's rivers, polluting them heavily. The situation in Dublin in the nineteenth century mirrored that of London, although Dublin's canal-based water system provided a better water supply than was available to many Londoners at the time.

In the mid-nineteenth century both Dublin and London suffered outbreaks of cholera. At this stage the causes of cholera and many other water contagious diseases were unknown. However, a paper by Bernard Mullins and M.B. Mullins published in the *Transactions of the Institution of Civil Engineers of Ireland* in 1848 makes it clear that although the underlying mechanism was a mystery, the link between cholera and inadequate drainage was well known.

In London the 'great stink' of 1858, when the Houses of Parliament were brought to a standstill by the overpowering stench from the River Thames, brought matters to a head. At that time it was widely believed that foul odours, the 'miasma', were the direct disease vector for cholera. The 'great stink' ultimately lead to sufficient funding being made available to allow Joseph Bazalgette to implement his London drainage plans. These plans, many years in preparation, involved a series of interceptor sewers that collected the wastewater before it discharged into the river. The foul water was transported many miles downstream of London to a pumping station where the wastewater was pumped to an outfall works.

In Dublin in 1857 control for the city's drainage was assigned to Parke Neville, the city's first full-time surveyor. By 1857 he had already published plans for a system of interceptor sewers for Dublin. This plan was redeveloped many times over the coming years with input from Bazalgette initially and in later years by Neville's successor, Spencer Harty, with input from Bazalgette's successor George Chatterton. Between 1851 and 1879 Parke Neville built 65 miles of new sewers and 30 miles of existing sewers were upgraded. Many miles of these new sewers comprised brick ovoid sections, 3ft 3in tall and 2ft wide. The ovoid section was chosen to maintain the velocity of flow when the volume of flow was low.

All this drainage work added to the quantity of wastewater and sewage being discharged into the River Liffey. Parke Neville's plans for interceptor sewers were delayed and alternative schemes involving the construction of additional weirs on the Liffey to increase aeration were considered and ultimately rejected. The construction of interceptor sewers was finally begun in 1896, ten years after Neville's death, and continued until 1906. This system involved interceptor sewers along significant sections of the river within the city boundaries. The northern interceptors were connected to the interceptor sewers on the south bank via a siphon under the river at Eden Quay. The southern interceptor extended to a pumping station in Ringsend, from where the sewage was pumped to settlement tanks at Pigeon House Harbour. Further work on Dublin's main drainage continued during the twentieth century with additional interceptor sewers, an 831ft-long 7ft-diameter tunnel under the Liffey at Ringsend and, most recently, a major treatment works at Ringsend, at which wastewater is also received from a number of satellite towns.

In Belfast, sewers previously discharged directly into the nearest watercourse and hence into the River Lagan and Belfast Lough. Underground brick-vault storage tanks, a screen-house and a pumping station were constructed between 1892 and 1896. The bulk of sludge in Northern Ireland was collected at local sewage treatment plants all over the area and brought by lorry in cake form, or by tanker in liquid form, to the Duncrue St treatment works on the northern shore of Belfast Lough. The resulting sludge was taken to sea and dumped in a designated area. In a similar manner, Dublin's sludge was also dumped at sea. However, in 1998, an EU directive prohibited the dumping of sludge at sea. In Belfast the sludge is incinerated and in Dublin it is pelletised for use as a fertiliser, although incineration is planned.

The most recent addition to Belfast's drainage system is a tunnel that collects storm water from all parts of the city, together with any sewage that overflows the weirs at times of severe storm, and transports it to the Duncrue St works, where it is treated before being discharged on an ebb tide.

DRAINAGE

Arterial drainage refers to the improvement of river flow for the purposes of land drainage and flood relief. Such drainage may occur in conjunction with navigation works. Interest in arterial drainage in Ireland dates back to the passage in 1715 of the Act of the Irish parliament 'to encourage the drainage and improvement of the bogs and unprofitable low-lying lands, together with the improvement of inland carriage and the dispatch of goods between one part of the kingdom and another.'

The Act had little effect and drainage was next championed in the early 1800s by the Dublin Society. The Society aimed to improve and develop Irish agriculture, and Thomas Prior, one of the Society's founders, published a treatise on 'a new method of draining marshy and boggy lands'. This interest in land drainage developed with John Foster, MP bringing forward a bill in 1809 'to appoint Commissioners to enquire and examine into the nature and extent of the several bogs in Ireland'. The work of the Bogs Commission, which extended from 1809 to 1813, led to surveys of the Irish bogs by Richard Griffith, William Bald, Alexander Nimmo, who wrote the drainage entry in *The Edinburgh Encyclopaedia*, and others.

This early work on drainage was designed to increase the area of arable land with the additional objective that the main drains were to be designed for use eventually as canals. The expense of the drainage works was to be borne by the landowners who benefited. This was a feature of many of the subsequent drainage initiatives, such as More O'Ferral's 1832 Act, Lynch's Committee of 1835, and the Drainage Acts of 1842 and 1863, and led to the most easily drained land being improved, and bogs, which were uneconomic to drain, being ignored.

Works on the River Shannon during the period 1839–1849 aimed to improve both the navigation and drainage of the river and were very successful. Prior to this work, stone beacons were built in some locations adjacent to the river to mark the location of the roads so that coaches could traverse them safely during times of flood. This scheme showed that greater benefits could be achieved by improving the outfall of rivers and streams, thereby improving the drainage of the adjacent land, than could be achieved by draining bogs.

In 1846 the Board of Works under John McMahon produced plans for combined drainage and navigation works to make the **Lower Bann (U7)** navigable from Lough Neagh to the sea, thereby reducing the level of Lough Neagh by approximately 6ft. The flow in the Lower Bann proved too fast for navigation and over the years many plans to abandon the navigation, to allow increased flow to alleviate flooding, were proposed and ignored. Eventually, additional drainage work was carried out in the 1930s.

Much of the land drainage of the nineteenth century was carried out under the 1842 arterial drainage Act. One reason that so much work was done under this Act was to give relief to the many thousands of poor who were employed on drainage works during the

Portna sluices on the Lower Bann, County Londonderry

Famine of 1845 to 1848. In 1847 the daily average number of men employed on drainage works was 20,000. At the same period the number of labourers working on relief works proper being supervised by the Board of Works, mainly roads, peaked at 734,000. For example, in County Clare almost 22 per cent of the total population were employed on relief works.

Putting poor, emaciated, unskilled persons to work on land drainage was considered appropriate, but it led to increased costs for the landowners who were responsible. This issue was eventually addressed, but the perceived inefficiency of the Board of Works led to a new Act in 1863, which put control of the design and execution of drainage schemes in the hands of the landowners. This move did not reduce the cost of drainage works.

The Land Purchase Acts of 1881 led to land being transferred from a relatively small number of significant landholders to a large number of their previous tenants. This was undoubtedly progressive but it had the secondary effect of making drainage works under the 1863 Act unworkable.

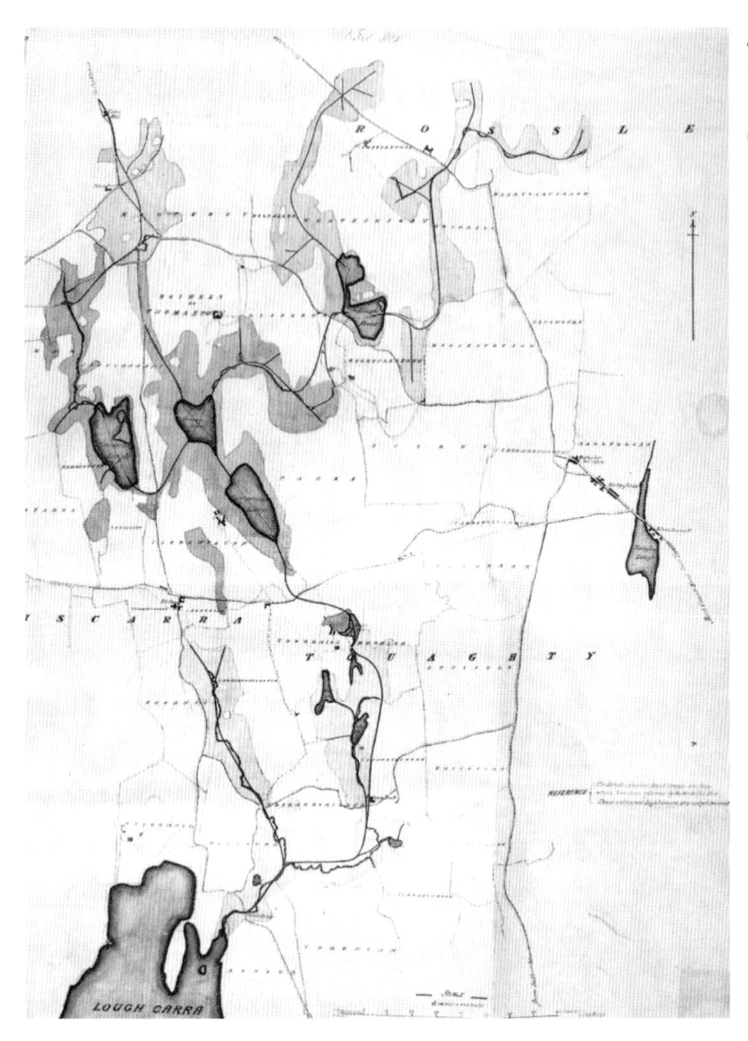

Arterial drainage proposals for the Lough Carra area of County Mayo

The funding mechanism, whereby the cost of the drainage works was borne by the landowners who benefited, was suited to local drainage works but was inappropriate for improving the flows in the lower sections of the principal rivers. The main arterial outflows were not improved, which limited the potential to drain land upstream, because allocating the cost to the landowners who ultimately benefited was not feasible. This severely limited the effectiveness of arterial drainage in Ireland until eventually, in 1925, the Irish Free State changed the funding model so that the state and local authorities undertook to cover the capital costs of arterial drainage works, leaving the landowners to contribute toward the maintenance costs.

Since 1925 the Office of Public Works (OPW) in the Republic and the Northern Ireland Rivers Agency have continued to undertake significant drainage and implement flood relief systems both separately and in cooperation. In the south, the Owenmore and Barrow

Drainage work on the River Boyne in County Meath OFFICE OF PUBLIC WORKS

schemes were implemented and in recent years many significant flood relief schemes have been and are being undertaken. The establishment of the OPW's hydrometric service, which records flows in Irish rivers, continued the Board of Work's scientific approach to flow prediction, a history which dates back to the work of its former chief engineer, Robert Manning, whose formula describes open channel flow.

POWER GENERATION

Before the development of the steam engine there were three sources of energy that could be used to power mechanisms for hoisting, grinding, cutting, or pumping water, namely wind, water, and animal or human power. The use of human power was well illustrated by the descriptions of pumping and hoisting mechanisms by George Semple in his *Treatise on Building in Water*, published in 1775, in which he describes the construction of Essex Bridge in Dublin.

Wind power in Ireland can be dated back to the thirteenth century, but its use was by no means common. However, there was a spate of wind-powered mill construction between 1770 and the end of the Napoleonic wars, particularly on the Ards and Lecale peninsulas in County Down. This activity was driven by an increase in the profitability of cereal

Blennerville Windmill near Tralee in County Kerry

production. The most notable surviving windmills are those at **Ballycopeland (B11)** in County Down, Skerries in north County Dublin and at Blennerville in County Kerry.

By far the most significant source of power in Ireland was waterpower. Water mills have a very long history and although many of Ireland's mills are now derelict, the many weirs on Ireland's rivers show how prevalent they once were. Water wheels can be divided into two types: vertical and horizontal. Evidence of the use of both forms in Ireland dates back to at least the seventh century. With a horizontal mill wheel, the flow from the millrace is funnelled into a chute that directs a jet of water onto the vanes of a horizontal water wheel, which lies below the mill wheel to which it is connected by a rigid shaft. The remains of many examples of this form of mill have been found, particularly in the Cork region, but amongst the most interesting examples is the tide mill at **Nendrum (U27)** in County Down. This mill was part of the monastic settlement of the Nendrum monastery and was in operation in the first half of the seventh century. The headwater was provided from an artificial enclosure that was filled by the tide.

By the eighteenth century the majority of mill wheels were vertical wheels. The vertical wheel is very ancient and was described by Vitruvius. Vertical wheels can be undershot, overshot, breast shot or pitch back, an efficient form of overshot wheel developed by the engineer John Smeaton. Undershot wheels are easy to construct and rely on the velocity of the flow impacting on the wheel. This form of waterwheel was improved by the engineer Jean-Victor Poncelet. An alternative improvement involved turning the undershot wheel into a breast-shot wheel. Breast-shot wheels take advantage of the potential energy due to the difference in the head upstream and downstream of the mill. The overshot wheel is the most efficient form of wheel. It was suitable for locations with a significant fall but limited flow because it makes optimum use of the water's potential energy. In an overshot wheel, the water flows over the wheel into buckets rather than flowing under the wheel. Further improvements to the basic wheel forms were introduced by engineers in the nineteenth century including John Rennie, Thomas Hewes and William Fairbairn. One of Fairbairn's suspension cast iron wheels is still in full working order at the Midleton distillery in County Cork.

There were many significant hydropower sites in Ireland during the eighteenth and nineteenth centuries. The Ballincollig Gunpowder Mills in County Cork was one of the largest hydropower sites in Europe. The mills had over 5½ miles of millraces powering up to thirty installations.

Ultimately, the water wheel evolved into the water turbine and later waterpower began to be used to generate electrical power. The story of hydroelectric power generation in Ireland is largely the story of the semi-state electricity supply company. The first hydroelectric power facility in Ireland was the Shannon Scheme at **Ardnacrusha (M29)**. This scheme was implemented in 1925 by the fledgling Irish Free State in cooperation with Siemens Schukert of Berlin. The scheme cost £5,200,000, took three years to build, and

Constructing the hydropower station at Ardnacrusha, County Clare ESB ARCHIVES

provided 85MW of electricity (later increased to 110MW), more than enough to power the entire country at the time. In 1937 a dam was constructed on the River Liffey at Pollaphuca, which created the reservoir at Blessington. Each of the two generators at the hydro station produce 15MW of electricity. Later two smaller turbines where installed on the River Liffey at Golden Falls and Leixlip, each of which produce 4MW. In 1946 the **Erne Scheme (U13)** at Ballyshannon in County Donegal was constructed. This includes two 10MW generators at Cliff and two 22.5MW generators at Cathaleen's Falls. The **Inniscarra (M30)** and Carrigadrohid dams on the River Lee provided a total of 27MW. Small generators were also constructed at Clady in County Donegal and at Parteen Weir upstream from Ardnacrusha.

The final Irish hydroelectric scheme was developed in 1974 at **Turlough Hill (L40)** near Glendalough in County Wicklow. This is a pumped storage system comprising two lakes connected by a tunnel. During off-peak times water is pumped up 287m from the lower reservoir to the upper reservoir. This water is released during times of peak demand thereby providing the grid with an additional 292MW.

The shifting balance between oil supply and demand, and concerns that fossil fuel consumption is driving climate change, has led to increased interest in traditional power sources. In recent years wind turbines have reappeared on the Irish landscape, with the Arklow Bank turbine field providing 25MW. Tidal power is also being reconsidered.

MARITIME STRUCTURES

Dublin Port DUBLIN PORT COMPANY

From the earliest of times Ireland, being an island, relied on its contacts with the rest of the world by sea, whether with its near neighbour or across oceans to far-off shores. Safe havens, developing into harbours and ports, became fundamental to its very existence.

The earliest visitors of note to Ireland, however unwelcome, were the Vikings, who brought with them the need for harbours and the ability to identify suitable sites on which to build sheltered havens for their craft. The present-day locations of Dublin and Limerick, among others, owe their position to some extent to these early raiders and their nautical requirements.

EARLY PORTS AND HARBOURS

Early sheltered havens were positioned to take full advantage of the natural landscape, river estuaries protected from the exposed coastline by protruding promontories, sheltered bays often shielded by islands – where ships could be anchored offshore or beached at low tides. These havens developed into settlements where goods, cattle and, indeed, people could be assembled to await shipment or dispersal to inland locations following arrival. The sailors determined, from experience, which inlets and sheltered areas were best suited to their needs and many of these locations developed over time into ports and harbours. Following the choice of an area that met the maritime requirements of the time, a settlement, usually at the head of a river estuary or at the lowest point of an available river crossing, developed, if the demands of the shipping fraternity could be met. Wharfs, quays, storage sheds and, most importantly, the ability to access these facilities at all states of the tide, provided probably the greatest challenges for those who designed and built them. The main difficulty with ports built on estuaries was the problem of silt and fine materials being brought downstream by the associated rivers, thus producing the need for dredging, particularly below the lowest tide level, to allow access at all times to the harbour walls. The early mariners, who considered such issues as tidal flows, prevailing winds and water depths, influenced the development of many harbour sites.

The early quays established by the Vikings were simply vertical timber posts driven into the bottom of the riverbanks to allow ships to moor and handle their cargoes on to the tops of the banks across horizontal timbers. The timber riverbank structures continued to develop up until the early twentieth century using hardwoods (such as greenheart and iroko), but always in rivers or sheltered docks where the action of wind and waves was not a problem. In the eighteenth and nineteenth centuries, the major problem was shallow water in the channels and berths caused by the silt-laden rivers slowing down as they flowed through the harbours and depositing their suspended materials forming mud or sand banks at the entrance.

SHIPPING DEMANDS

As deeper berths were required, quays had to be built further out into the sea to form not only a deeper-water facility to load and unload cargo and passengers, but also to increase the number of berths available in a harbour. As vessels changed and developed from small wind-driven craft to larger mechanically propelled ships, and from manual loading and unloading to mechanised methods, and as they grew larger and even more sophisticated, the harbours and ports had to be developed in tandem, otherwise commercial advantages would have been lost at both national and international levels.

The designers and builders of ships soon realised that the longer the ship's waterline the faster it could be driven through the water. Larger ships also produced better commercial returns for their owners in, for example, fewer crews and lower pilotage costs. This knowledge led to larger ships, taller masts with greater areas of sail, deeper draughts and a larger capacity for cargo/passengers. Not only did ship development lead to the need for longer quays, but more significantly from a civil engineering point of view, deeper water alongside the quays and their approaches. Harbours, however, followed the development of ships more slowly than the shipyards that produced bigger and more modern vessels, and often the ships could not be utilised to their full potential in many of the ports, not only in Ireland, but worldwide, until the harbour authorities had completed the 'catch-up' development.

Inner Dock, Dublin, prior to redevelopment

Early in the eighteenth century it became clear that the larger individual harbours could not exist without national or, at least, external funding. Harbour Boards, Ballast Offices, Harbour Commissioners or local Chambers of Commerce were established to oversee the development of ports on a strong commercial footing in Belfast, Dublin, Limerick, Waterford and Cork. By the late nineteenth and early twentieth centuries, many harbours on the east coast of Ireland were driven commercially by the attractions of the relatively short sea crossing to Britain and the more sheltered aspects of the Irish Sea. Harbours on the west coast of the island were developed to meet mainly the needs of the fishing fleets, and for communication with the offshore islands. In addition, transatlantic shipping opted to use the deep-water facilities being developed along the south coast, particularly in Cork Harbour.

By the end of the eighteenth century, many of the major Irish ports had begun building 'wet' docks in which a vessel could remain afloat at all stages of the tide. Wet docks are controlled by lock gates that allow vessels to enter the dock from the river or estuary at high tide and then retain the water at a constant level whilst the tide outside the gate ebbs away. This system has two advantages, one being the ability to keep the ship afloat whilst in port, therefore not straining the hull, the second being the convenience of having the ship and the quay level remaining constant for the loading and unloading facilities, a task that had previously been very labour intensive. Visitors to Dublin can see good examples of wet docks from the eighteenth and nineteenth centuries on either side of the River Liffey. Towards the end of the nineteenth century the major ports in Ireland had developed some deep-water berths alongside new or refurbished quaysides to meet the then current demands. The earlier sites, those based on river estuaries, were extended out towards the sea to enhance the capacity of a port and these were mainly developed using masonry walls constructed on either side of the harbour approach. Nevertheless the long arms extending outwards did not negate the need for dredging and this remains an expensive operation up to the present day. Without a deep-water berth attached to the land-based transport infrastructure, vessels often had to anchor offshore and unload their cargoes into lighters (small barge-type craft) to complete the journey into the harbour. This double handling was costly and time consuming, and harbour authorities were conscious that their 'traffic' might go elsewhere if their own facilities were not improved to accept the larger vessels. Until the development of dredging equipment in the mid-nineteenth century, it was very difficult to maintain deep-water channels up to the quays, particularly in estuary-based harbours.

DRY (GRAVING) DOCKS

Ports, such as Dublin, Cork, Waterford, and later Belfast, developed shipbuilding and/or repair facilities that brought with them the need for dry or graving docks – a facility where

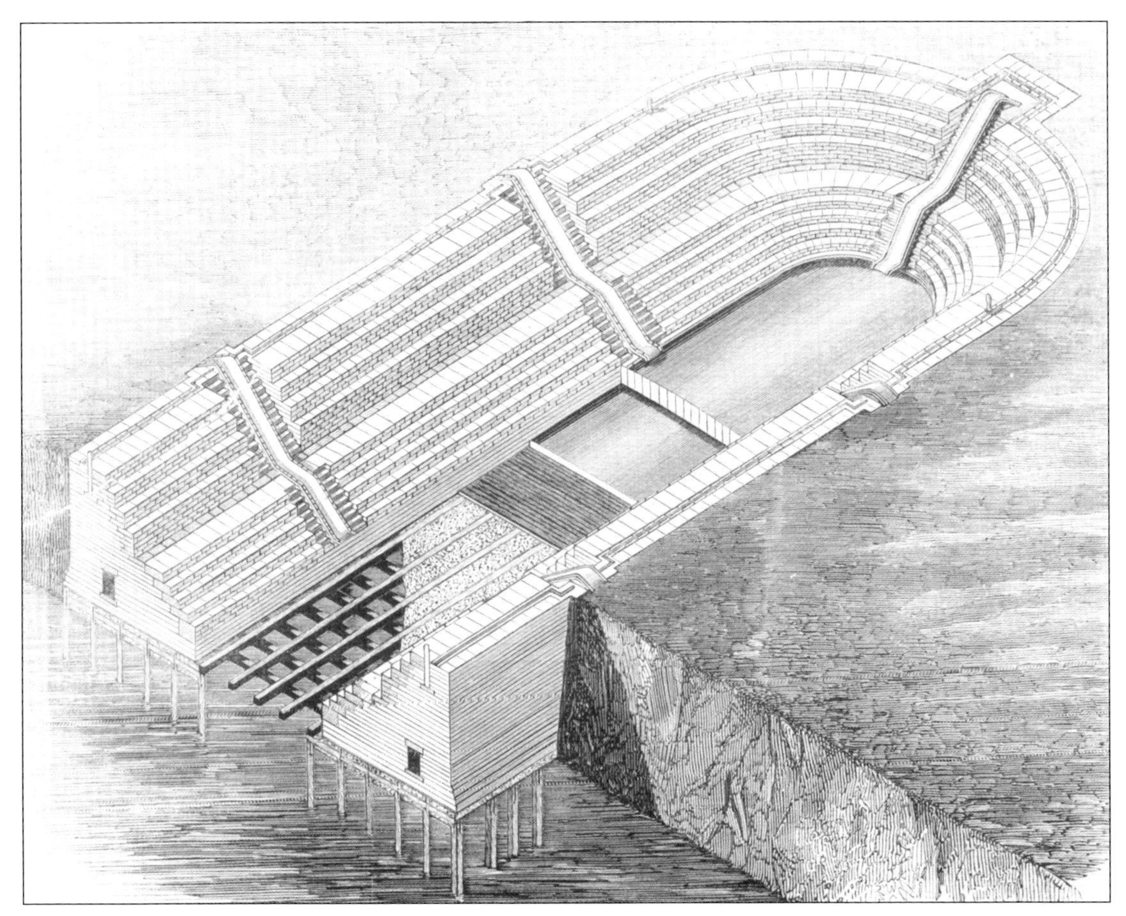

Dublin No.1 Graving Dock

ships could be 'beached' safely, in a working environment, to have their underwater areas repaired and painted. Graving docks are large masonry structures that allow vessels to enter at high tide; lock-type gates close after the ship enters and allow the tide outside to ebb. As soon as the ship is propped off the sidewalls, the water in which it floats is pumped or drained out, and the vessel is effectively left on dry ground. The lower hull and its fitments, rudder and propellers can then be cleaned, serviced and painted before the dock is flooded to refloat the vessel and the gates opened at high tide, thereby releasing the vessel back into the sea.

These docks are often very complex civil engineering construction projects, since they are usually situated on poor ground due to their proximity to the estuary or river. Designs had to allow for the weight of a ship (or the dock full of water, which is actually heavier) without significant settlement, which would fracture the pump connections. During cleaning of the empty dock, the gate would be shut and the dock would be pumped dry producing an uplift from the surrounding groundwater, particularly at times of high tide, which effectively tried to make the dock 'float'. Traditionally dry docks were constructed in masonry, which relied on self-weight to hold them down in the 'empty' scenario, the stepped sides being solid stone ballast. Modern dry docks are constructed with reinforced concrete and the design includes anchors down into the bedrock using either reinforced concrete piles or rock anchors.

The earlier graving docks had lock-type gates, but as larger docks were required, especially for the shipbuilding industry, a floating gate was constructed which was pumped out and floated to one side as the ship entered the dock, then floated across the entrance into grooves in the masonry/concrete wall. The gate was then flooded to at least high-tide level, thus acting as a heavy sunken vessel to form the gate. The dock could then be pumped dry. Some good examples are still visible to the public in Belfast's 'Titanic Quarter' (**B1**) and at **Limerick** (**M3**), opened in 1873.

In ports with large areas of land available nearer to deeper water, slipways were also built. In shipbuilding areas, such as Belfast, Derry and Cork, slipways were constructed to enable the hulls of large ships to be built in the 'dry'. When the hull was completed it was launched down the gradual slope of the slipway, sliding over greased baulks of timber, into the water. The floating hull would then be fitted out alongside a quay with all its internal decks, machinery and equipment. On completion, the new vessel would then be dry-docked in a graving dock for final painting and adjustments to the underwater areas, prior to its first or maiden voyage. Very large slipways can still be seen in Belfast.

In smaller ports, mainly fishing harbours, concrete slipways were built with cast-in railway tracks set to a gauge to suit bogeys constructed to carry the vessel. The bogey is

Dock cranes

let down into the water at high tide, the vessel positioned over the bogey, and both are pulled up the slope by winches, the vessel gradually settling onto the bogey. A variant of this is a large mobile mechanism that lifts the vessel out of the water and traverses it to a hard standing area where work can be carried out to its underside. Vessels can also be housed in tall purpose-built sheds, for example at Killybegs in County Donegal, where larger deep-sea fishing boats can be handled by all of the above methods. As the ships became larger, more complex building docks developed (very large graving docks) where ships were constructed or assembled completely inside the dock before it was flooded, releasing the new vessel to the sea. A large building dock is still in use in Belfast.

PORT ACCESS AND FACILITIES

The establishment of any port could only progress if the infrastructural backup, which enabled goods and produce from the hinterland to be transported to it, developed with it. Transportation of agricultural goods was the Irish economy's backbone, as produce and livestock could not be stored indefinitely in areas remote from the sea ports.

Initially, canals were built, following to a large extent the gentle slopes of the rivers and loughs, down to the harbours. Later, and in many cases along the same routes as the canals, the railways connected the inland market towns to the nearest port, which brought about the demise of the canals. Railways flourished during the nineteenth century (see Chapter 3), but it was not until the early twentieth century that heavy-capacity goods vehicles were developed, together with concrete or asphalt surfaces on wider roads, thus allowing faster transport from the source (not the canal or rail depot) to the docks, which in turn resulted in the carriage of freight by railways in Ireland virtually to disappear during the second half of the twentieth century. On the other hand, passenger trains appear to be making a recovery; for example, the rail link between Limerick and Galway has recently been restored.

Methods of material handling at docks have had to alter with time and will continue to develop during this century. Labour-intensive handling has given way to the highly mechanised movement of goods, integrated with land-based transportation systems. In the main harbour areas, this can be broken down into various types of goods requiring specialised handling equipment:

- Bulk dry goods, e.g. cereals, coal, and quarry products
- Bulk wet goods, e.g. fuel oil
- 'Loose' goods, e.g. vehicles, steel, plant, and timber
- Containerised goods – lift-on/lift-off (lo-lo)
- Vehicular traffic – roll-on/roll-off (ro-ro)

Not all Irish ports can handle all of these commodities, but most ports can cater for several – some specialise in ro-ro only, for example, Larne, Dun Laoghaire and Rosslare. The largest ports – Belfast, Cork and Dublin – can cater for all types of goods.

During the second half of the twentieth century, specialist harbours and quays were built to supply oil and power industries where ultra-deep-water facilities were required, usually away from the established ports, for example in Bantry Bay (Whiddy Island) and Belfast Lough (Kilroot).

With the ever-changing developments in shipping, harbours have had to upgrade their berthing arrangements for each new generation of ships. These comprise fast twin-hulled passenger and vehicle-carrying vessels with double decks, which require special 'tidal' ramps and docking equipment unique to the specific ship; mono-hulled modern craft needing ramps capable of operating at all states of the tide, and capable of accepting the vessel's docking bow or stern to the ramps. Larger ships need deeper berths, larger and heavier cranes, faster bulk unloading equipment and larger warehousing and passenger handling facilities. The continuous development of harbours and ports can be attributed to the skills of the civil engineer who strives to keep abreast of the advancements in the shipbuilding industry.

DUBLIN HARBOUR AND PORT

From medieval times, Dublin, and somewhat later Belfast, have developed into the two largest ports on the island, each being the capital city of their own jurisdictions. The problem that beset Dublin Harbour from early times was the silt being deposited into Dublin Bay by three rivers: the Dodder, the Liffey and the Tolka. The rivers flowed over the strands at low tides, producing natural channels, but they were constantly shifting with the action of the tides and wind.

In medieval times, the harbour was mainly on the south side of River Liffey, and much further upstream, nearer to Christ Church Cathedral. As part of the eighteenth century development the quays were extended down to the junctions of the rivers Dodder and Liffey, near where the early fishing village of Ringsend evolved, and now known as Sir John Rogerson's Quay, constructed between 1717 and 1727. It was in this area that the **Grand Canal Circular Line (D6)** (completed in 1796) entered the River Liffey near to the mouth of the River Dodder, an L-shaped wet dock being constructed there by the canal company. There were three tidal locks of varying capacity and three graving docks. The latter were all subsequently filled in, but the L-shaped dock is still operational, one of the tidal locks (Buckingham Lock) has been restored, and one of the graving docks is currently being restored. The Grand Canal docks did not prove to be commercially viable, mainly due to the narrowness of the tidal locks and deposits of silt from the River Dodder at the entrance. Compounding these problems, or because of them, the Ballast Board decided to construct

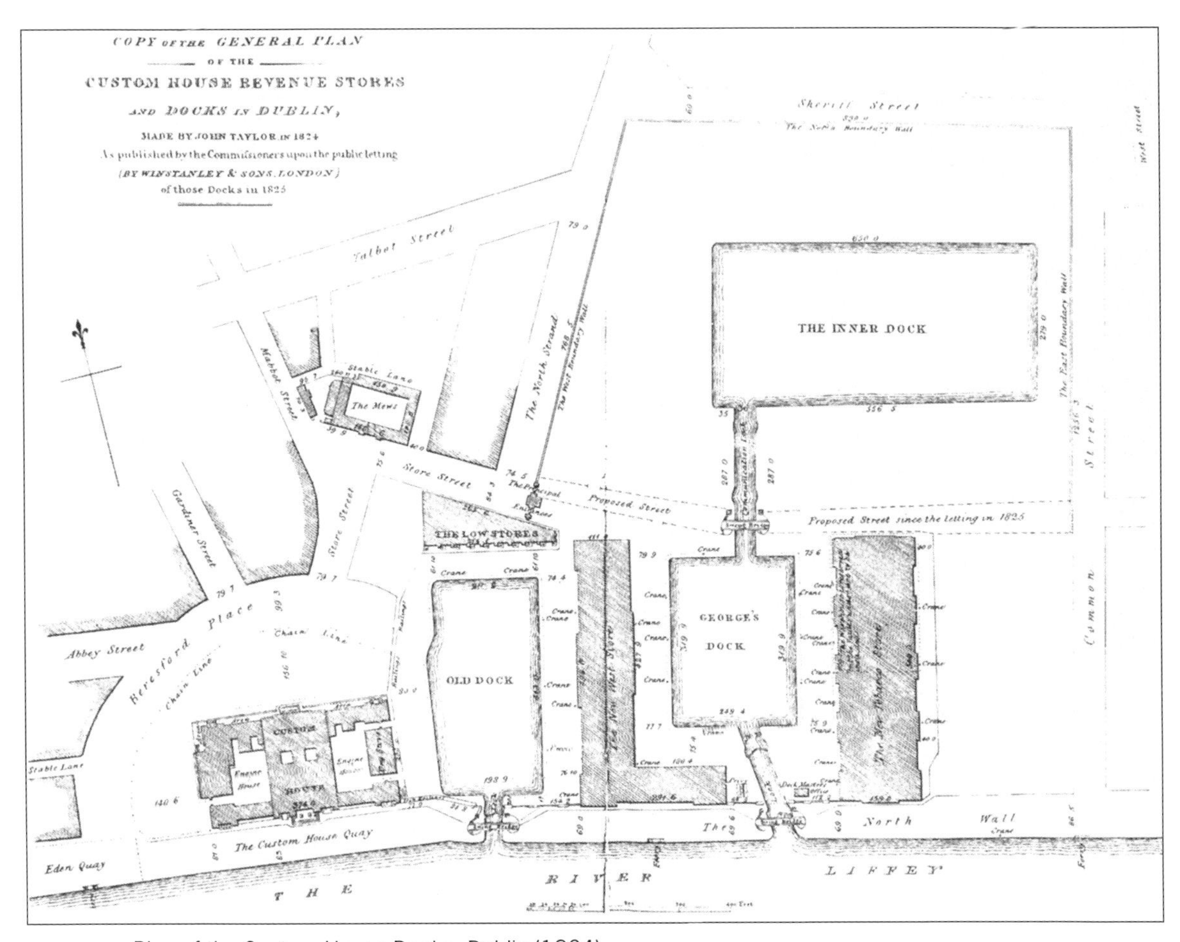

Plan of the Custom House Docks, Dublin (1824)

wet docks further upstream on the north bank of the River Liffey. These docks were developed adjacent to the then new Custom House, and were Old Dock (1796), George's Dock (1821) and Inner Dock (1824). These docks were entered through a tidal lock, the gates being designed, after Rennie's death in 1821, by Thomas Telford, who also completed the Inner Dock. As deeper quays were developed further east, the so-called **Custom House Docks (D3)** fell into disuse, although the docksides and lifting bridges can still be seen today, the area having been redeveloped for offices and apartments. By 1860, the No.1 graving dock (410ft long by 70ft wide) was completed in Dublin port by William Dargan, and still exists, the present gate having been renewed in 1931, but recently the dock has been filled in. Construction of the North Wall quay extension in 1871–85 formed the Alexandra Basin. The wet dock (or basin) was built under the direction of the Dublin port engineer, Bindon Blood Stoney (1828–1909), using large monolithic concrete blocks, some weighing up to 350 tons, cast on shore and transported into position by a large floating

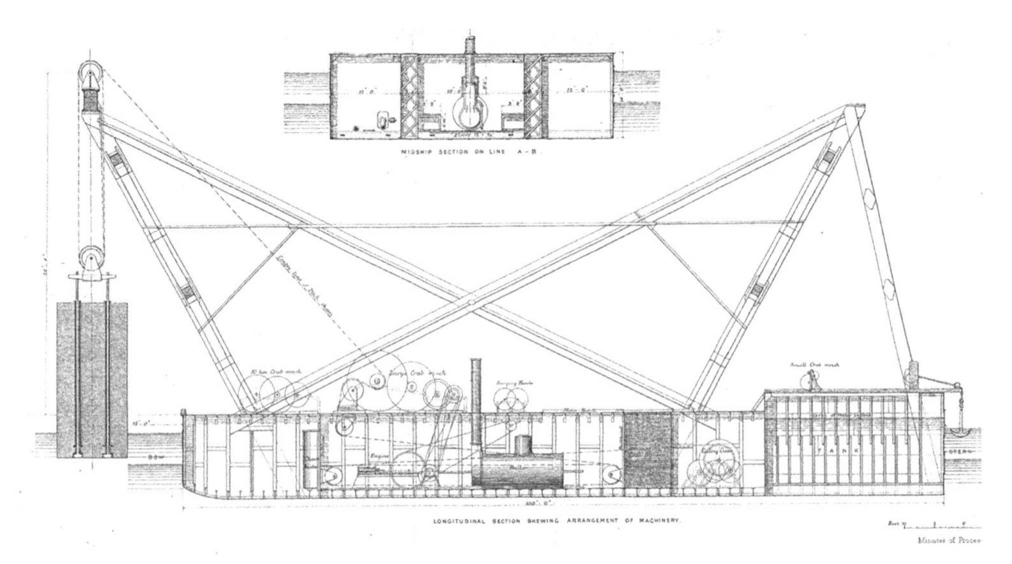

Stoney's block-laying float

crane. The foundations for these blocks were prepared in some 30ft depth of water using a diving bell, which has been preserved. The Alexandra Basin was opened in 1885.

Up until the nineteenth century it was generally accepted that building piers out into the sea on the line of the rivers would increase the depth of water at the quay, allowing larger ships to use a port. To some extent this worked, but in Dublin the amount of silt being brought down by the three rivers formed banks of sand and silt at the end of the wall that blocked or formed a 'bar' at the entrance to the port.

In the early eighteenth century, work began to build a straight wall, replacing an earlier bank on the south side of the harbour entrance, initially with timber caissons (boxes) floated into place and filled with stones and sand – the start of the South Bull Wall. By the end of the eighteenth century, when the wall was finally completed, it was one of the longest sea walls in Europe, and was completed using twin parallel masonry walls infilled with sand. The original timber construction between Ringsend and Pigeon House Fort was rebuilt in masonry around the middle of the eighteenth century.

However, the sand bar and shifting channel beyond the end of the South Bull Wall were still causing problems. Several prominent engineers and surveyors – including William Chapman, John Rennie and William Bligh (Captain Bligh of HMS *Bounty* fame) – were commissioned to consider a solution. In the early part of the nineteenth century, they recommended that the problem be addressed by increasing the flow of water out of the harbour, thus scouring the riverbed.

It was Francis Giles (1787–1847) and George Halpin (1779–1854) who, in 1819, suggested that a further breakwater should be constructed from Clontarf on the northern side of Dublin Bay to a position opposite the Poolbeg Lighthouse at the end of the South Bull Wall. The **North Bull Wall (D1)** was completed by 1825 in limestone and granite masonry, forming a new mouth for the River Liffey. The design of the wall resulted in the

Entrance to Howth Harbour

tidal and river flows contributing to the scouring of the sand bar at the entrance leading to a considerable improvement in the depth of the shipping channel leading to the port and Sir John Purser Griffith, the Port Engineer, was moved to remark that this was 'a noble example of directing the great sources of power in nature for the use and convenience of man'.

By 1895, suction dredgers, which replaced bucket dredgers, were in use in the harbour to straighten and deepen the channels, within both the harbour and its immediate approaches. By the mid-twentieth century, ro-ro ferries, both conventional and high-speed services, were well established and containerised cargo has become the largest part of the traffic handled, as the port has been developed to meet modern transportation developments. The port has now moved further eastwards to deeper 'all-tide' quays and wharfs almost 3km beyond the location of its early beginnings. Warehousing requirements and all the paraphernalia of docklands had to travel eastwards with the ever-deepening berths, but many of the earlier structures still survive. Whilst rail links have diminished, new road connections, including a 4.5km-long twin-bore tunnel, feed heavy traffic under the northern suburbs to connect with the motorway system.

Modern road bridges and footbridges over the river complement the earlier bridges and the area has been redeveloped to incorporate a pleasant riverside walk, complete with artistic pavement furniture.

When 'artificial' harbours were constructed in areas away from river estuaries to give safe havens, mainly for fishing vessels, they did not necessarily suffer from silting, but had to have stout breakwaters to shelter quays and needed to be capable of withstanding the ravages of bad weather. Prior to the completion of the North Bull Wall in 1825, the mail packet steamers from Britain were encountering difficulties with the shallow harbour and the lack of facilities, often causing the ships to unload both mail and passengers into small boats to be rowed ashore. In 1807, it was decided to build an artificial harbour at **Howth** (**D8**), north of Dublin. In 1818, the new harbour was being used as the mail packet harbour for Dublin, but Howth had access difficulties that, in 1826, caused the service to be transferred to Kingstown (Dun Laoghaire). The construction of **Kingstown Harbour (D9)** commenced in 1817. John Rennie senior designed two embracing arms later known as the east and west piers. Following his death, his son (later Sir) John Rennie – both father and son had a great influence on the design of harbours in Ireland – designed small wings at the end of the piers, but the final form of these was eventually completed by the Board of Public Works in 1833. Fast ferries operating from Holyhead in Wales continue to use this harbour.

All three ports were developed in the first quarter of the nineteenth century, with John Rennie and his son having a prominent civil engineering input to their design and construction. Howth developed mainly as a fishing port, and later the provision of major marina facilities has helped it and Dun Laoghaire maintain their maritime presence. Dun Laoghaire has also continued to develop as a ferry port, adjusting and amending its facilities to ensure commercial stability, mainly based around ro-ro passenger and freight services.

PORT OF BELFAST

Belfast was established between the rivers Owenvarnagh (Blackstaff) and the Farset where they joined the River Lagan on its north bank. The ships anchored in the area adjacent to the mouth of the River Farset and the High Street Quays were established in 1670. These were subsequently closed over in 1757 by the development of High Street. Most of the harbour development that followed was based on the County Antrim (north) side of the River Lagan and the quays were built there from the River Blackstaff over the next 200 years. Chichester Quay ran from Church Street to Ann Street (1769–1830), and was built by Thomas Greg. Donegall Quay (built in sections) ran from Ann Street to Queen's Square at the end of High Street (1712–1822). Queen's Square to Albert Square was constructed from 1783 to 1838, and Albert Square to Corporation Street from 1804 to1838, the final section being completed by Henry Joy Tomb.

The Irish parliament passed legislation in 1785 to form a Ballast Board, which oversaw the construction of the quays, and by 1790 Belfast had become the third largest port in Ireland. Docks, constructed at right angles to the River Lagan at Queen's Square, Albert Square and Corporation Square are now all filled in, and the Farset River has been culverted under the present High Street.

Shipbuilding started on the County Antrim side of the River Lagan in 1791 with William Ritchie's shipyard in the Limekiln Docks in the area of High Street and by 1800 the first dry dock, Clarendon No. 1, was in use. By 1812, forty ships had been delivered from these yards and by 1820 the first steamship in Ireland was built. A second, larger dry dock was constructed and opened in 1826 – Clarendon No. 2 – and in 1838 an iron steamer was launched by Kirwan & McCune.

The Belfast Dock

Both dry docks are preserved near the Belfast Harbour Office. The outer tidal dock, the **Clarendon Dock (B1)**, was completed in 1840. Further east towards Belfast Lough, the York Dock was completed by 1830 and in the same year the first steam dredger was purchased. Shipbuilding commenced on the County Down side in 1853. In 1847 the Belfast Harbour Act was passed, forming the Belfast Harbour Commissioners, who were given wide-reaching powers enabling them to reclaim land, build new quays and docks and also dredge the channel seawards, and still maintain these roles today.

In the 1830s William Dargan proposed straightening the course of the River Lagan from below the mouth of the Farset towards the open lough. The first stage of this work was completed by 1839 and spoil from the dredging placed on the shallow County Down side of the River Lagan to form what became known as Dargan's Island. Following the second phase of straightening, this island grew in size and became known as the Queen's Island, becoming the home of the main shipbuilding activities of Harland and Wolff.

Engineering growth in Belfast developed in parallel with the early port, producing machines mainly for the linen spinning and weaving industries. These industries developed a workforce in ironworking and machinery skills, and these in turn complemented the burgeoning shipbuilding industry. The first foundry in Belfast (Hill Street) was opened in 1741, and by 1800 three more were in production.

The expansion of the County Antrim shipyards was undoubtedly due to the harbour improvements and the developing trade links, especially with Europe and America.

The low tidal range in Belfast encouraged the development of dry docks (graving docks) to be constructed directly off the River Lagan. As lands to the County Down side were reclaimed, the Belfast Harbour Commissioners built dry docks, in 1867 the **Hamilton Dock (B1)** (to the rear of the Odyssey complex), in 1889 the **Alexandra Dock (B1)** and in 1911 the **Thompson Dock (B1)**. All these docks are still intact and the associated Pump

House between the Alexandra and Thompson Docks has been refurbished and is used as a visitor interpretative centre and cafe. The only dock without its gate, and therefore tidal, is the Alexandra Dock – currently the home of HMS *Caroline* – a First World War fast cruiser, which up until December 2009 was home to the Royal Naval Reserve in Northern Ireland, and is now decommissioned.

The shipyard complex of Harland and Wolff eventually covered the entire area on the County Down side of the River Lagan, bounded in part by an aircraft manufacturer (now Bombardier), the Connswater River, and the transport routes to the north of the county.

In 1968 a large, new dry dock, Belfast Dock, was built to facilitate even larger craft, and in 1970 a new building dock was opened to construct large super tankers. Both of these docks are still in use today, serving much reduced shipbuilding activities. The large area now vacated by the shipyard, which for many years was the largest and most innovative in the world, is now in the hands of commercial developers and home to the 'Titanic Quarter'.

The commercial harbour at Belfast has stayed mainly on the County Antrim or north side of the river and is capable of handling all types of cargo and passenger traffic.

It would be misrepresenting the importance of other ports and harbours around the island to concentrate on Dublin and Belfast alone, indeed over the last three to four centuries, some of the smaller harbours were for a time more important, whilst Belfast in particular is a relative newcomer.

OTHER PORTS

Cork harbour and its inlets – considered by some to be the second largest harbour in the world – was a very important port in the eighteenth and nineteenth centuries and served as the victualling base for the British Navy's Atlantic fleet. Many of the army and naval defences, barracks and quays can still be seen. Many of the large passenger liners made the sheltered waters of the Cork Harbour area their first or last stop on their transatlantic voyages.

In the north of the island, Derry city on Lough Foyle, the foundation of which dates back to the seventh century had, by the nineteenth century, developed into a major port and shipbuilding centre. Derry saw many of the emigrant ships leave for the New World, and also developed commercial trade with Britain. The port has now moved to deeper water at Lisahally, utilising the quays left following the closure of large industrial complexes.

Larne harbour at the mouth of Larne Lough dates back to medieval times, and today concentrates on ro-ro ferries to Britain.

Newry, at the head of Carlingford Lough, developed as a port on the completion of the **Newry Canal (U21)**, which connected it to Lough Neagh. A sea lock and channel some 1½ miles long was constructed to make Newry accessible at all stages of the tide.

Facing page: Harbour Light at Dunmore East, County Waterford

Today, the port activities have moved to deeper waters at Warrenpoint on the east shore of Carlingford Lough where they now concentrate on ro-ro and lo-lo traffic.

Rosslare Harbour on the southeast coast was developed by the railway companies to provide access to and from Wales, and has developed ro-ro ferry traffic to south Wales and France.

Killybegs in County Donegal is now the largest deep-sea fishing port in Ireland, developed from a sheltered haven used by local craft since the seventeenth century.

Like earlier harbours, Galway and the Shannon Estuary ports and harbours, and Bantry Bay with its twentieth-century oil industry terminal facilities, have all developed or declined with the ebb and flow of export and import demands. Some have been allowed to decay and silt up, others have developed leisure facilities, such as marinas.

LIGHTHOUSES

Ireland has over 3,000km of coastline. Lighthouses are placed where the land meets the sea, on rocky outcrops or isolated islands just offshore, and predominantly on approaches to harbours and sea lanes near to the shore. Like lifeboats, lighthouses come into their own on dark stormy nights when navigation and seamanship are tested to their limit as vessels approach the coast and harbours. Nowadays, navigational aids, such as global positioning systems (GPS), provide reassuring support.

The earliest lighthouse in Britain and Ireland was built in the fifth century on **Hook Head (L35)** at the entrance to Waterford Harbour, by a Welsh monk who maintained a fire in a metal basket close by his residence. A light has been maintained, albeit intermittently, since that time, with the present tower dating from 1245. A second light was set up at Youghal (1190) by the nuns of St Anne's convent, and during the seventeenth century a group of bonfire-type lights were established at Charles Fort and the Old Head of Kinsale, also in County Cork, Loop Head in County Clare and the Baily Lighthouse on Howth Head. In 1711 a similar structure was built on one of the Copeland Islands (Lighthouse Island) off Donaghadee, County Down. It was during the nineteenth century that most of Ireland's lighthouses were constructed.

Dundalk Pile Light, in Dundalk Bay, County Louth COMMISSIONERS OF IRISH LIGHTS

There are now over eighty lighthouses in Ireland, maintained by the Commissioners of Irish Lights (CIL), who are responsible for all Irish

Duncannon Lighthouse, Waterford Harbour

lighthouses and navigation buoys around the coastline of Ireland. The CIL obtain their income from ships entering Irish ports, depending on their gross tonnage, with fishing boats and small pleasure craft generally exempted.

The earliest light, produced by bonfires, fuelled by coal, peat or wood, was usually placed on high ground or stone towers, and was somewhat weather dependent and often suffered from 'blanketing' of the firelight by its own smoke. Near the end of the eighteenth century it became clear that a lens and possible reflector would be needed to project light from candles or oil wicks a greater distance, and if installed inside a structure with a glass-sided top would be less weather dependent. In 1789, a large plano-convex lens reflector was built into the existing structure at the Baily on Howth Head by Thomas Rogers. He was also responsible for altering other lighthouses from candles to oil wicks. He built new structures, the oldest and most significant historically being the wave-washed lighthouse on the **South Rock (U23)** off the County Down coast. This structure still survives, although derelict, and was replaced by a lightship in 1877, and subsequently by a modern navigation buoy. Rogers is credited with the development of the revolving light, allowing each light to have a different characteristic, which helps ships to identify specific lights – rather like a 'fingerprint' for each light. This identification system applies to this day.

At the beginning of the nineteenth century, the Dublin Ballast Board appointed George Halpin senior as Inspector of Lighthouses for the whole of Ireland, and during his tenure almost forty lighthouses were constructed or started, which included some of the most spectacular structures including the first Fastnet and Tuskar lights.

In 1898, a new **Fastnet Rock (M2)** lighthouse was designed and constructed in granite by William Douglass. It was first lit in 1904 and is perhaps the most iconic of all the Irish lighthouses.

Where it was impossible to build structures because of poor ground or shallow water, specially designed lightships were anchored in permanent positions and had individual characteristics similar to lighthouses.

The power to produce light was firstly fire followed by candles, oil wick (1780), gas (1815), paraffin – vapour incandescent burners (1904) and electricity circa 1920.

The electricity was either generated on station or by mains supply (both with standby systems); one had an experimental isotope generator, and currently more use is being made of solar power stored for nocturnal use.

Historically, three types of lighthouse accommodation were built for the keepers: those with land access, although often in remote areas with purpose-built housing nearby; those with living accommodation built into or around the light tower on isolated positions such as offshore islands or those with difficult access by land; and some lighthouses were serviced by watchkeepers from nearby towns and harbours. All Irish lighthouses are now unmanned, with some of the existing housing used as summer holiday accommodation. The lights are visited frequently by maintenance teams and controlled remotely from the CIL headquarters at Dun Laoghaire.

In whatever period one chooses to consider the construction of harbours, ports and their infrastructure, viz: lighthouses, quays, transport systems serving the coast such as canals, road networks and railways, all have been designed and constructed by civil engineers, many of whom were giants of their time and could be regarded today as having helped to lay the foundations of the modern society in which we live.

GAZETTEER

This section is a descriptive list of civil engineering works, not only of those referred to in the previous chapters, but other sites further illustrating the contribution of the civil engineer to human progress over many years.

The sites are arranged by province, except for those in the areas of Dublin and Belfast, which are treated separately. In each case, a map shows the approximate location of each site. For each site the following information is given:

- The location
- The national grid reference
- A brief description of the work
- The registered HEW number

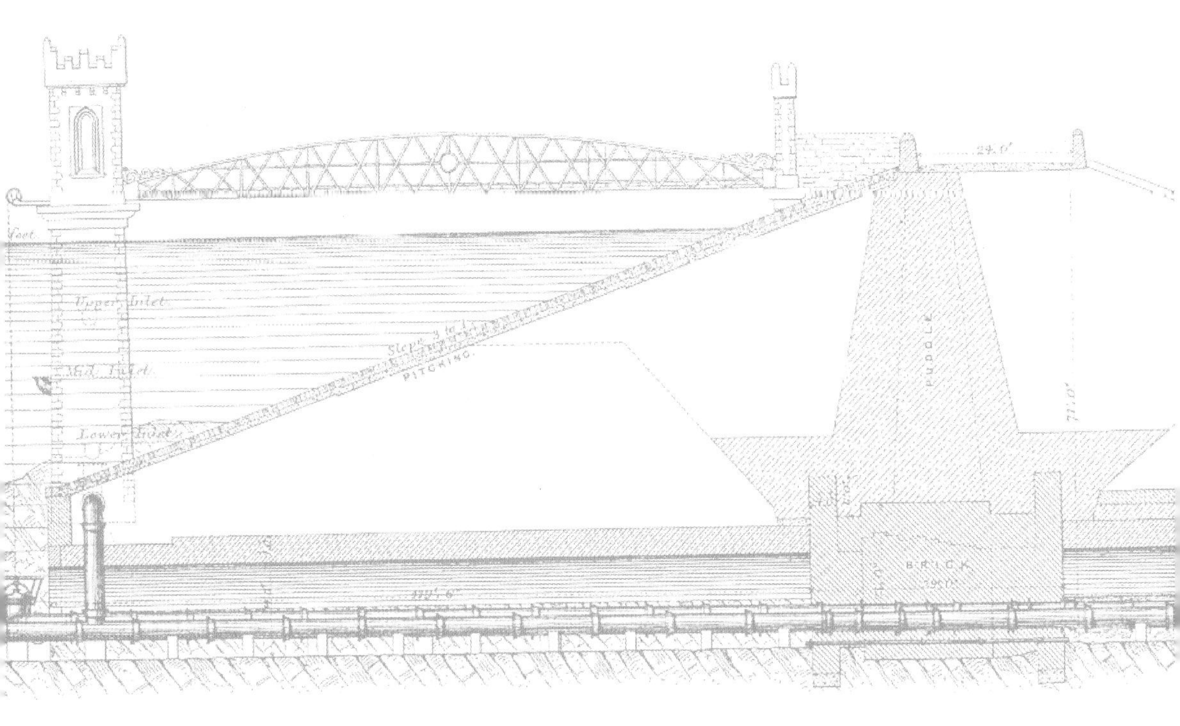

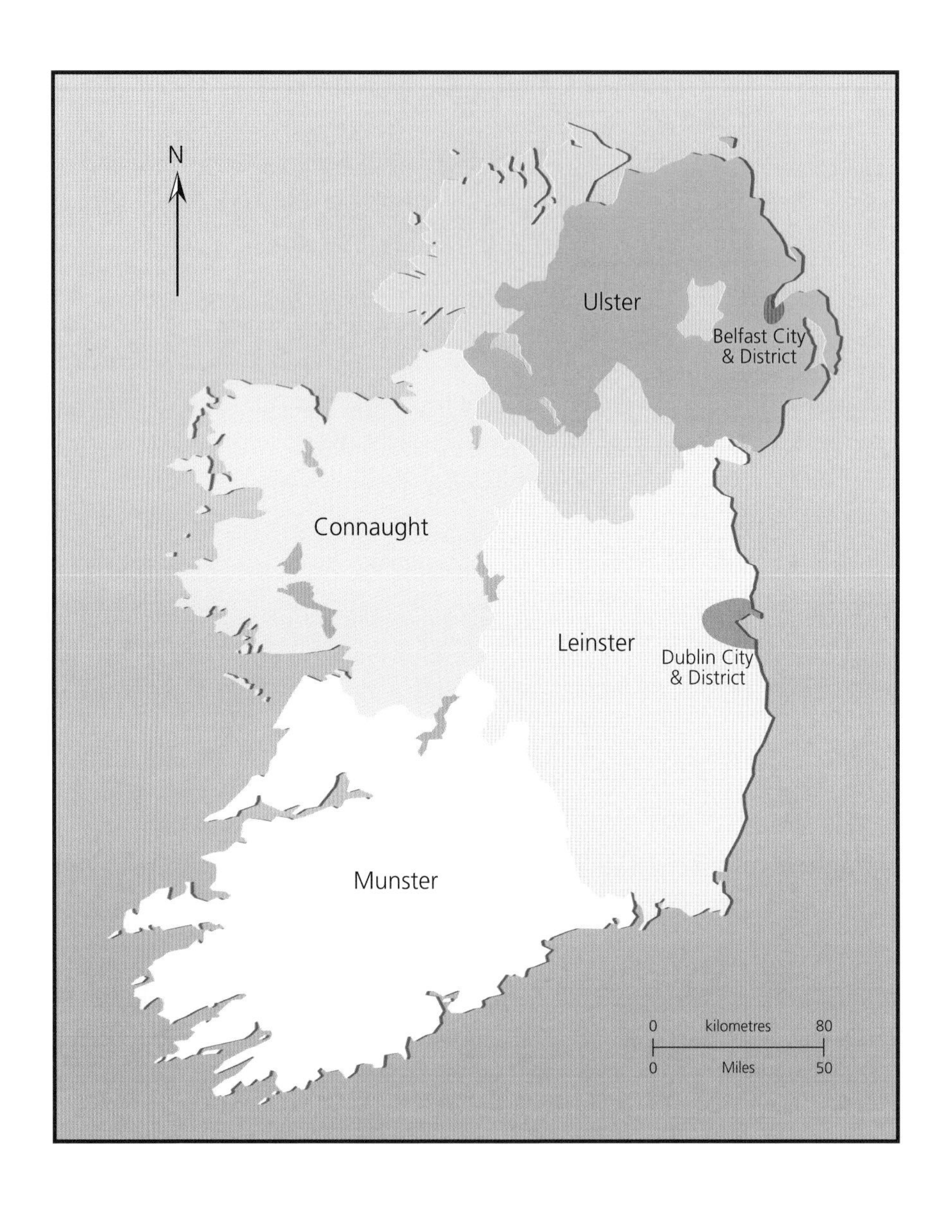
N
Ulster
Belfast City
& District
Connaught
Leinster
Dublin City
& District
Munster
0 kilometres 80
0 Miles 50

LEINSTER

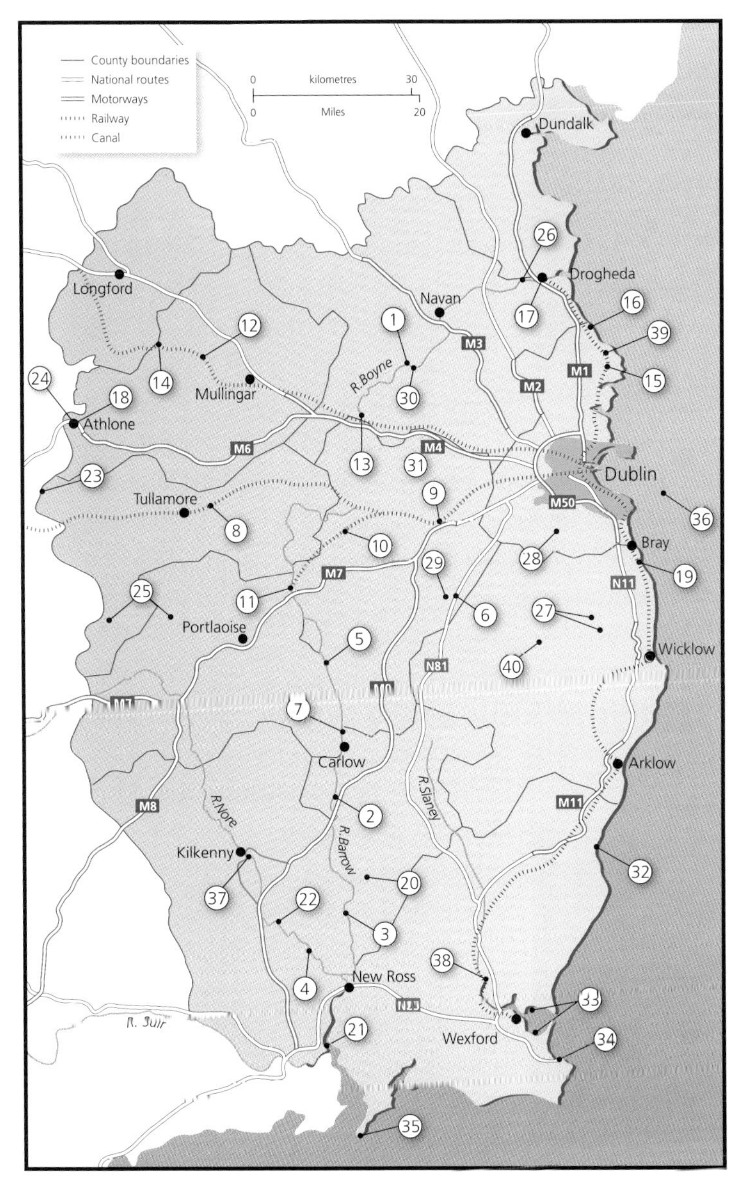

L1 Trim Bridge
L2 Leighlin Bridge
L3 Graiguenamanagh Bridge
L4 Inistioge Bridge
L5 Athy Bridge
L6 Pollaphuca Bridge
L7 Oak Park Bridge
L8 Grand Canal (Main Line)
L9 Leinster Aqueduct
L10 Grand Canal (Barrow Line and Barrow Navigation)
L11 Barrow Aqueduct
L12 Royal Canal
L13 Boyne Aqueduct
L14 Whitworth Aqueduct
L15 Dublin & Drogheda Railway
L16 Balbriggan Rail Viaduct
L17 Boyne Rail Viaduct and Bridge
L18 Shannon Rail Bridge (Athlone)
L19 Bray Head Tunnels
L20 Borris Rail Viaduct
L21 Barrow Rail Viaduct
L22 Nore Rail Viaduct (Thomastown)
L23 Shannonbridge
L24 Athlone Road Bridge, Weir and Locks
L25 Birr and Kinnity Suspension Bridges
L26 Obelisk Bridge
L27 Vartry Reservoirs
L28 Bohernabreena Reservoirs
L29 Dublin Water Supply (Blessington)
L30 Trim Water Tower
L31 Straffan Suspension Bridge
L32 Courtown Harbour
L33 Wexford Harbour Improvement
L34 Rosslare Harbour
L35 Hook Head Lighthouse
L36 Kish Bank Lighthouse
L37 St John's Bridge, Kilkenny
L38 The Deeps' Bridge, Killurin
L39 Skerries Windmills
L40 Turlough Hill Pumped Storage Station

Trim Bridge

L1. Trim Bridge (N 801570), over the River Boyne at Trim in County Meath, consists of four pointed segmental masonry arches of 16ft span with a rise of 6ft 9in. The arch rings are formed of roughly trimmed, rectangular stones of varying thickness (3in to 6in). The 8ft-thick piers and abutments are founded on rock, and the bridge is 21ft wide with solid parapets finished with coping stones. The present bridge was most likely built in 1393 and is probably the oldest unaltered bridge in existence today in Ireland. During the 1970s, the Office of Public Works carried out an arterial drainage scheme, which involved lowering the bed of the river at this point by 4ft. As with many other bridges subject to the scouring action of rivers, the base of each pier of the bridge was surrounded with a reinforced concrete skirt, incorporating pointed cutwaters to assist the flow of floodwaters under the bridge. (HEW 3157)

Leighlin Bridge

L2. Leighlin Bridge (S 691654). It is generally accepted that the first stone bridge to be built across the River Barrow at this site was erected by Maurice Jakis, a canon of Kildare Cathedral, in or about 1320. It seems likely that Leighlin Bridge was rebuilt in the mid-seventeenth century, and widened in 1789. The present bridge consists of seven segmental masonry arches of varying span and rise, the largest being just under 30ft. The rise-to-span ratio is around 0.3 and the span-to-pier thicknesses are in the ratio of 5 to 1. In 1976 a cantilevered reinforced concrete pathway was constructed on the upstream face and the whole bridge pressure grouted and the arch soffits gunited. A bypass was opened in 1986, thus relieving the old bridge of the vibrations and heavy loads imposed by modern commercial traffic. (HEW 3142)

Graiguenamanagh Bridge

L3. Graiguenamanagh Bridge (S 708438). One of a number of aesthetically very pleasing bridges spanning the River Barrow, the present multi-span masonry arch bridge at Graiguenamanagh was erected in the 1760s to replace an earlier timber structure. The bridge, built in the Palladian style, was probably designed by George Smith, a pupil of George Semple, the architect responsible for Essex (now Grattan) Bridge over the River Liffey in Dublin. The spandrels reduce in size towards the ends of the bridge. The arch spans vary from 19ft 4in nearest the riverbank to 31ft 10in at the centre of the bridge, the shape of the arches being segmental with a rise-to-span ratio of 0.4. The arch rings are rusticated using local slaty stone. The piers vary between 7ft 2in and 5ft 10in in thickness. The internal construction of the spandrels is similar to Labelye's work at Westminster Bridge in London, i.e. longitudinal and cross-walls of rubble dividing the space into compartments filled with gravel. Westminster Bridge was the first major bridge in Britain in the eighteenth century, and Smith's work shows an awareness of contemporary practice. (HEW 3136)

L4. Inistioge Bridge (S 636377), spanning the River Nore at Inistioge in County Kilkenny, is a rebuilding of an earlier structure, which was severely damaged in the great flood of 1763. There are nine semicircular arches of 24ft 3in span and the width between the parapets is 16ft 9in. The rebuilding was carried out under the direction of George Smith. Smith applied the concept of Mylne's design for Blackfriars Bridge in London to the downstream face of the bridge at Inistioge. It is actually more true to the triumphal bridge model, having all nine arches equal and semicircular, resulting in a parapet that is horizontal rather than curved. The spandrels are of good dark-coloured rubble decorated with pairs of Ionic pilasters in pale and sharp-edged granite. (HEW 3165)

Inistioge Bridge (detail)

L5. Athy Bridge (S 681940). There has been a bridge at Athy since the early fifteenth century, when it was associated with a nearby fortress. A plaque, mounted in a pediment in the parapet over the crown of the central arch on the downstream side, records the fact that the present bridge over the River Barrow at Athy was built in 1796 by Sir James Delehunty, the Knight of the Trowel (i.e. a master mason). The bridge is known locally as Croomaboo, the name coming from the war cry of the Desmonds. This Palladian-style bridge, sponsored by Robert, Duke of Leinster, has five segmental arches, the central span being about 36ft with a rise of 7ft 10in. The width is a generous 30ft between parapets and there are flanking footpaths. The design was most likely by the engineers of the Barrow Navigation, who carried out works in the area up until 1812. (HEW 3139)

L6. Pollaphuca Bridge (N 957085). The design of this spectacular single-span pointed segmental arch bridge in 1820 is generally attributed to the Scottish civil engineer Alexander Nimmo. It carries the Dublin–Blessington road over the River Liffey at Pollaphuca on the Kildare/Wicklow border. The arch springs from rocks on either side of a gorge through which the river flows out of the Blessington Lakes. Over the arch, which is ornamented with a bold moulding, runs a cornice, above which is a solid parapet. There are tower abutments with blind and cross-loops and with battlemented parapets surmounting a richly recessed blocking. The spandrels have Gothic niches with hood mouldings. The arch span is 65ft, with a rise of 39ft 4in, and the maximum height of the bridge above the river is 150ft. Pollaphuca Bridge takes its name from the 'puca's pool', a deep pool in the river below a waterfall where legend has it that the puca (a mythical Irish spirit) lived. (HEW 3125)

L7. Oak Park Bridge (S 736798) near Carlow is an estate bridge cast at Colebrookedale in Shropshire by John Windsor and erected in 1818. The single span consists of five parallel cast-iron arch ribs of 40ft span and 6ft rise. Each rib is formed of two cast sections bolted together at the crown. The arch spandrels are enjoined rings decreasing in diameter from the abutment to the crown. Each half-rib is cast in one piece, the bottom segmental arched and the top horizontal girder sections being pierced with small holes of constant diameter in order to reduce the dead load. The inner ribs are of similar but somewhat simpler design to the outer ribs. The 15ft-wide deck is formed of iron plates spanning transversely across the tops of the ribs, which are braced transversely at intervals. The iron railings forming the parapets are comprised of vertical bars fixed between top and bottom rails, the bottom rail being supported on ball feet. There is a cast snake motif at regular intervals along the parapet railings. The abutments are of ashlar granite and extend 20ft on either side of the clear span. Cheaper limestone was used in the core of the abutments. This listed structure is maintained by the Agricultural Institute, whose headquarters are at Oak Park. (HEW 3275)

Athy Bridge

Pollaphuca Bridge

Oak Park Bridge near Carlow

L8. Grand Canal (Main Line) (O 177345 to N 033189). Work started in 1757 to connect Dublin by canal with the Shannon Navigation (**C1/C2**) via the Grand Canal. Thomas Omer surveyed the route and, following the incorporation of the Grand Canal Company in 1772, work was continued by John Trail, and later Charles Tarrant. The first section

Grand Canal 29th Lock FRED HAMOND

from the Dublin terminus at City Basin (St James's Street) to Sallins was opened to traffic in 1779. The canal crosses the River Liffey by the Leinster Aqueduct (**L9**) and continues to its summit level near Lowtown in County Kildare, some 280ft above low water at Dublin. Near Lowtown, the canal obtains a water supply from the Milltown and Blackwood feeders. Omer had planned that the canal would cut straight across the Bog of Allen, but John Smeaton advocated a more northerly route towards Edenderry, which was adopted. Smeaton had argued with William Chapman about the desirability or otherwise of draining the bog prior to cutting the canal. Smeaton's view prevailed and the canal was constructed at the same level as the existing surface of the bog, without allowing for a period of drainage. The result was that the land on either side drained into the canal and subsided, leaving the canal confined by high embankments. These have proved to be a constant source of trouble to the canal: major breaches of the banks occurred soon after completion, again in 1916, and more recently in 1975.

Richard Evans and his assistant John Killaly progressed slowly with much difficulty, and the line was opened to Daingean by 1797 and to Tullamore in 1798. Tullamore Harbour became the temporary terminus of the canal whilst it was decided how best to reach the Shannon. The route eventually chosen followed the valley of the River Brosna, joining the River Shannon just north of Banagher. Richard Griffith senior, the father of Sir Richard Griffith, supervised the work with John Killaly as his engineer and the bog was drained before this final section of the canal was built. The canal was finally opened to the Shannon in 1804. In 1950 the Grand Canal Company was merged with Córas Iompair Éireann and in 1986, responsibility for inland waterways passed to the Office of Public Works, and finally to Waterways Ireland.

The average width of the canal is 30ft and the average depth at the centre is 5ft with minimum headroom under the bridges of 8ft 6in at the waterline. There are five aqueducts, and dry docks at Tullamore and Shannon Harbour. Of the thirty-six locks, thirty are single and six double (to conserve water). The average width of the locks and the navigation under bridges is about 15ft. The shortest lock is 69ft 8in, the longest 88ft 7in, the narrowest 13ft 6in, and the widest 16ft 2in. Branches of the canal were completed to Athy (the Barrow Line), and to Edenderry, Kilbeggan and Ballinasloe. The branch to Naas and onwards to Corbally (known as the Kildare Line) was acquired in 1808 from the County of Kildare Canal Company. On this short canal, William Jessop, in 1789, developed his method of constructing skewed masonry arch bridges. The Grand Canal was connected to the River Liffey in 1796 by the construction of the Circular Line (**D6**) from the First Lock at Inchicore to the Grand Canal Dock (**D7**) near Ringsend. (HEW 3001)

L9. Leinster Aqueduct (N 876228) carries the Grand Canal over the River Liffey to the west of Sallins in County Kildare. It was designed by Richard Evans, the engineer to the Grand Canal Company, and construction work commenced in 1780. The aqueduct, in ashlar limestone masonry, has five arches, each of 25ft span, with a rise of 7ft 6in, and has pointed cutwaters. The overall length is 233ft, the width of the river at this point being 152ft. There is a 10ft 9in wide towpath on the north side, a 15ft 6in wide roadway on the

Leinster Aqueduct

south side, and a canal waterway of 16ft 6in width. Including the 18in-thick parapet walls, the overall width of the aqueduct is 46ft 3in. Another aqueduct, with two small arches, lies some 100ft further west of the end of the main aqueduct. There is provision for excess water to overfall from the canal to the River Liffey below and, in times of water shortage, the process may be reversed using temporary pumping facilities. (HEW 3096)

L10. Grand Canal (Barrow Line) and Barrow Navigation (N 778254 to S 682935 and S 682935 to S 725380). Between 1783 and 1791, the Grand Canal Company built a branch from the main line near Lowtown to the River Barrow at Athy in County Kildare, a distance of 28 miles. The descent of approximately 100ft from the summit level at Lowtown required nine locks and a small number of aqueducts – the principal one (**L11**), of three 40ft spans, carrying the canal across the River Barrow at Monasterevin in County Kildare. There was also a further branch of 10 miles to Mountmellick via Portarlington, but this is now derelict.

The designers involved were principally Richard Evans and Archibald Millar. The contractors for the section from Lowtown to Monasterevin were Thomas Black and others, and John MacMahon completed the remainder to Athy. The Barrow Navigation consists of 42 miles of river navigation, 11 miles of which are artificial cuts or canalised sections, the longest being the Levitstown cut of 2 miles between Athy and Carlow. There are twenty-three single-chambered locks between Athy and the sea lock at St Mullins at the tidal limit of the river. The locks for the most part are 80ft long by 16ft 6in wide. A total of twenty-two weirs along the river maintain water levels upstream of the canalised sections. Work commenced on improving the Barrow Navigation as early as 1715, but was not completed until 1791. (HEW 3003 and HEW 3004)

Athy Locks on the Barrow Line of the Grand Canal

The Barrow Line of the Grand Canal crossing the River Barrow near Monasterevin

L11. The Barrow Aqueduct (N622105) was designed by Hamilton Killaly (a son of John Killaly) and built between 1827 and 1831 by Henry, Mullins and MacMahon. It consists of three segmental masonry arches, each of 41ft 6in span, with a rise of 7ft 6in, giving a low rise-to-span ratio of 0.18. The arch voussoirs are rusticated ashlar limestone and the garlanded decoration over each river pier gives the aqueduct an attractive overall appearance. The canal over the aqueduct is 17ft wide, with an 8ft-wide towpath on either side. The overall length of the aqueduct is 207ft, of which 134ft 6in is over the waterway. At the eastern end of the aqueduct is a small lifting bridge and nearby was a group of canal warehouses, now converted to apartments. (HEW 3127)

L12. Royal Canal (O 173345 to N 060760). The Royal Canal Company was incorporated in 1789 to build a canal to link Dublin with the River Shannon at Cloondara in County Longford. Work commenced in 1790 on the main line, but progress was slow and it eventually took twenty years to complete the 90 mile link. The Royal Canal, unlike its rival Grand Canal, did not offer passenger and freight services between Dublin and Limerick and had to be content with serving the smaller centres of population in the north Midlands and the upper Shannon.

The line had originally been surveyed by John Brownrigg, the engineer to the Directors General of Inland Navigation, but the direction of the actual route and the detailed

Royal Canal bridge plaque

engineering design became the responsibility of the engineer to the canal company, Richard Evans. However, he died in 1802, by which time the canal was open to the nineteenth milestone from Dublin. He was succeeded as engineer by John Rennie senior, whose masonry bridges from Blackshade Bridge, west of the Boyne, to Coolnahay are of noticeably higher quality than Evans' work on the earlier section of the canal. They were, however, more expensive, and by 1809 the company was unable to progress further. As a result of the financial difficulties the construction of the canal was taken over in 1813 by the Directors General of Inland Navigation and was completed to the River Shannon by 1817 under the direction of their engineer, John Killaly. The contractors were David Henry, Bernard Mullins and John MacMahon, destined to become one of the leading contracting firms in Ireland in the early nineteenth century.

The canal commences at a sea lock at the North Wall quays in Dublin and passes through the Spencer Dock, built in 1875 by the Midland Great Western Railway (MGWR). The canal is carried over the Rye water near Leixlip in County Kildare on a substantial embankment, the river flowing underneath in a 230ft-long tunnel, 30ft wide with a semicircular arched roof. Major aqueducts were constructed at Longwood in County Meath over the River Boyne (**L13**) and at Abbeyshrule in County Longford over the River Inny (**L14**), and smaller ones at a number of other locations. The summit level of the canal ends at Coolnahay, some 6 miles beyond Mullingar, which was reached in 1806. There are a

Boyne Aqueduct on the Royal Canal

total of forty-six locks on the canal, ten of these being double-chambered. The canal company was purchased in 1845 by the MGWR, who built their line alongside the canal from a Dublin terminus at Broadstone to a few miles west of Mullingar. Ownership of the canal transferred from Córas Iompair Éireann to the Office of Public Works in 1986 and a vigorous programme of restoration was begun. Now under the control of Waterways Ireland, the navigation has been restored and was reopened from Dublin to the Shannon in 2010. There is public access to much of the towpath, as is the case with the Grand Canal. (HEW 3002)

L13. Boyne Aqueduct (N 692453). The aqueduct was completed in 1804 and carries the Royal Canal over the River Boyne near Longwood in County Meath. The river at this point is about 40ft wide and flows through the central arch of the aqueduct. The designer, Richard Evans, was faced with the problem of providing a structure capable of withstanding not only the forces applied by the canal waters, canal traffic and the dead weight of the masonry but also the considerable lateral forces resulting from the river when in flood. The resulting viaduct in rusticated ashlar limestone masonry has three arches spanning the river and its flood plain. The arch profiles are each 40ft span with a rise of about 16ft. The voussoirs in the arch rings are 2ft deep and the tops of the spandrel walls terminate in a cornice running the length of the aqueduct. The 6ft-thick piers have a simple architectural feature rising from curved cutwaters. The massive abutments are splayed at the base for stability to a width of 107ft 6in. This stretch of the canal is about 35ft wide and narrows to 20ft over the aqueduct, the total length of which is 132ft. There are 8ft 9in wide towpaths on each side of the waterway. The main Dublin–Galway railway line runs alongside the canal and crosses the river on a parallel masonry arch viaduct, built in 1849, which has three flat elliptical arches of rise-to-span ratio of around 0.25 and of similar span to those of the aqueduct. A smaller single arch aqueduct of 28ft span over the R160 is located just to the east of the river crossing. (HEW 3128)

L14. Whitworth Aqueduct (N 232600) carries the Royal Canal over the River Inny near Abbeyshrule in County Longford. It was designed by the talented Irish canal engineer John Killaly, and constructed between 1814 and 1817 by the partnership of Henry, Mullins and MacMahon. The aqueduct consists of five segmental masonry arches, each spanning 20ft with a rise of 5ft. The width of the structure at the springing level of the arches is 38ft 9in, but this reduces to 35ft at the level of the canal. The arch extrados is formed of cut limestone voussoirs that project from the face of the spandrel walls. The river piers and splayed abutments are carried up from river level in ashlar limestone detailing to a cornice. The spandrel walls are of rough-hewn coursed limestone, as are the parapet walls, which extend between decorative panels over the piers. The overall length of the aqueduct is 165ft

Whitworth Aqueduct carrying the Royal Canal over the River Inny

and the overall width 35ft at the canal level. There are towpaths 8ft wide on both sides of the waterway, which is 15ft 6in at the waterline reducing to 13ft at the bottom. (HEW 3124)

L15. Dublin & Drogheda Railway (O 166350 to O 100748). A route between Dublin and Drogheda was surveyed by the English engineer William Cubitt, and work commenced in 1838 at Kilbarrack on the section between the Royal Canal and Portmarnock. As no gradient was to exceed l in 160, in order to accommodate the locomotives of the day, some heavy engineering was required, such as the cuttings at Malahide and Skerries and the embankment at Clontarf near Dublin. In February 1840, John Macneill was appointed engineer to the Dublin and Drogheda Railway Company and a new Bill was expertly steered through parliament by Daniel O'Connell. A new contract was awarded to Messrs Jeffs, who completed the Killester cutting and Clontarf embankment. The remainder of the line to Drogheda was contracted to William Dargan and William McCormick. The line became part of the Great Northern Railway (Ireland) network in 1876.

The line was opened on 24 May 1844 by the Lord Lieutenant and Macneill received a knighthood for his services to the community. Amiens Street (now Connolly) Station was completed two years later and is at a high level, the tracks being carried northwards on a viaduct of seventy-five arches as far as the Royal Canal. To cross the canal, Macneill used iron-latticed girders of 144ft span, which were the largest of their type in the world at the time of erection. These lasted until 1862, when increasing loads required an intermediate pier to be erected in the bed of the canal. The latticed girders were replaced by Pratt-type trusses in 1912. The original timber viaduct at Malahide was replaced in 1860 by a wrought-iron structure and this in turn between 1966 and 1968 by the present concrete viaduct. In 2009, the viaduct suffered a partial collapse when one of the piers was destroyed by tidal scour, necessitating the replacement of two spans. By comparison, the masonry

Drogheda Station, 1844

viaduct at Balbriggan (**L16**) is as originally constructed by William Dargan in 1843. The other viaducts over the river estuaries at Rogerstown, Gormanstown and Laytown were also originally in timber, but were replaced in wrought iron in the 1880s, and were further strengthened in 1996 as part of the upgrading of the Dublin–Belfast rail link. (HEW 3098)

L16. Balbriggan Rail Viaduct (O 202640). At Balbriggan, at the northern extremity of County Dublin (Fingal) between the town and a small fishing harbour, the Dublin and Drogheda Railway is carried over four roads and a small river on an eleven-arch viaduct built in 1844 by the contractor William Dargan to the design of John Macneill. Each of the segmental arches is of 30ft span with a rise of 10ft. The viaduct comprises arch rings with internal brick spandrel walls spanning between 6ft-thick brick piers, the outer arch rings and spandrel walls being of limestone. The individual voussoirs are 3ft thick and 2ft deep. The total length of the viaduct between the 10ft-thick abutments is 390ft. The rails are carried on 8in-thick spandrel covers supported on seven diaphragm walls, which are alternately 15in and 12in thick. Footpaths, each 6ft 6in wide, are carried at rail level on each side of the viaduct on cast-iron arches spanning between the 41ft-wide stone-faced brick piers. Cast-iron pilaster caps over each pier are connected by iron railings along the length of the viaduct. (HEW 3035)

Balbriggan Rail Viaduct

Boyne Viaduct and Bridge, 1932

L17. Boyne Rail Viaduct and Bridge (O 098754). The imposing viaduct and bridge over the estuary of the River Boyne at Drogheda, about 32 miles north of Dublin (O 098 754), was constructed in its original form between 1851 and 1855 as the final section of the Dublin and Belfast Junction Railway, linking the Dublin and Drogheda Railway at Drogheda with the Ulster Railway at Portadown. John Macneill, having introduced the iron-latticed girder bridge to these islands in 1843, suggested that such a bridge would be suitable for the main span of the Boyne crossing, but left the details of the design to James Barton, chief engineer to the railway company. The latticed girders were designed to be continuous across the supports. The principles of multi-lattice construction in wrought iron were first applied on a large scale in the Boyne Viaduct. This led to a significant increase in knowledge of the structural properties of wrought iron and encouraged more exact computation of the stresses in structural components. The design for the viaduct consisted of three latticed girders crossing over the main river channel, two of 141ft and one of 267ft between bearings, and fifteen semicircular masonry arch spans, twelve on the south side and three on the north side of the river, each of 60ft clear span. The massive masonry piers are composed mainly of local limestone and are founded on rock (except for pier 14). Initially, the contractor was William Evans of Cambridge, who had raised the tubular girders at the Conwy Bridge on the Chester & Holyhead Railway. However, he went bankrupt on the Boyne Viaduct contract following difficulties in finding a firm foundation for pier 14. The work was completed in 1855 by the railway company under the direction of Barton and a young resident engineer, Bindon Blood Stoney.

Increasing axle loads led in 1932 to the replacement of the latticed girders with simply supported mild-steel girders, the central span being given a curved top chord to improve its appearance. To reduce the maximum load on the bridge, the rail tracks were interlaced, thus allowing only one train at a time across the bridge. Designed by George Howden, chief engineer of the Great Northern Railway (Ireland), the present structure was erected by Motherwell Bridge and Engineering. (HEW 3013)

Shannon Rail Bridge, Athlone

L18. Shannon Rail Bridge (Athlone) (N 036420). The Midland Great Western Railway Company (MGWR) was the first to reach Athlone on the banks of the River Shannon. In order to proceed to Galway, then being considered as a possible transatlantic port, George Willoughby Hemans designed a substantial bridge to carry the railway across the river. The bridge, completed by Fox Henderson & Company in 1851, is formed from twin wrought-iron latticed bowstring girders, 20ft 4in deep at their centres and spanning 176ft, a central cantilevered opening span of 117ft 10in and two side spans, one of 45ft 1in on the east and one of 54ft 7in on the west side of the river. The girders are carried on piers consisting of pairs of 10ft-diameter cast-iron cylinders spaced 27ft 11in apart and cross-braced. The overall length of the bridge is a little over 580ft. The side spans of the bridge were renewed in steel troughing in 1927 and the opening span was fixed in the closed position in 1972. The track was originally double across the bridge, but was later converted to single line working. (HEW 3079)

L19. Bray Head Tunnels. (O 285173 to O 285147) The line around Bray Head, between Bray and Greystones, passes through a series of tunnels. In 1855, Isambard Kingdom Brunel designed three tunnels for double track, but only single track was laid. Built by William Dargan, the tunnels were the Brabazon (630ft), Brandy Hole (900ft) and Cable

Bray Head tunnels

Rock (429ft). The rocks of Bray Head are Cambrian quartzite and very hard, making tunnelling expensive, but they are also very unstable.

The Brabazon Tunnel was bypassed in 1876 by two short tunnels. Two timber trestle viaducts, to the south of Brandy Hole, were replaced by a single masonry arch and two arched sections built to support the cliff above the track. A further deviation became necessary in 1889, including another bridge replacement, and the Brandy Hole and Cable Rock tunnels were relined with brick and stone without altering the loading gauge. Further erosion at the base of the cliffs necessitated a final major deviation. The Long Tunnel, designed by C.E. Moore, was commenced in 1913 by Naylor Brothers of Huddersfield and completed in 1917 by direct labour under the resident engineer W.H. Hinde, after Naylor had withdrawn from the contract. The tunnel is 3,252 long, 15ft in internal height and has a 9ft 6in-diameter ventilating shaft 100ft deep. (IHEW 3123)

L20. Borris Rail Viaduct (S 730500). In 1854 the Bagenalstown and Wexford Railway was incorporated in a bid to reach Wexford by an inland route. Although the route selected ran through wild and unremunerative countryside, there was some prospect of through traffic. By the end of 1858 the standard gauge line had been completed from the junction with the Great Southern & Western Railway at Bagenalstown as far as Borris in County Carlow. The line was completed to Ballywilliam by 1862, by which time the company had run out of funds and was declared bankrupt. Links with Wexford (via Macmine Junction) and New Ross (via Palace East Junction) were eventually completed, but by then the Dublin,

Borris Rail Viaduct

Wicklow & Wexford Railway had reached Wexford. Passenger services were terminated in 1931, but occasional freight trains worked the line up until closure in 1963.

To the south of Borris Station, the line was carried over the Mountain River valley on a substantial limestone viaduct of sixteen spans. The semicircular arches are each 35ft diameter and are carried on slightly tapering piers some 10 to 15ft above the average ground level. The piers are 5ft thick and 18ft wide at the base. The viaduct passes over a public roadway at the town end and has a total length of about 470ft. The engineer to the railway company was William Le Fanu and the main contractor for the line was John Bagnall. There is pedestrian access to the viaduct as part of a nature trail. (HEW 3110)

Barrow Rail Viaduct

L21. Barrow Rail Viaduct. (S 683147) A new rail route across the south of County Wexford, between Rosslare Harbour and Waterford, necessitated the construction of a number of bridges, viaducts and tunnels. The major construction project was the crossing of the estuary of the River Barrow.

Designed by Sir Benjamin Baker, who was also involved in the design of the famous Forth Rail Bridge in Scotland, the Barrow Bridge is, at 2,132ft, the longest rail bridge spanning entirely across water in Ireland and was, at the time of its completion in 1906, the third longest in either Britain or Ireland. The engineer to the GS&WR at the time was James Otway. At the west end of the bridge the railway enters Snow Hill tunnel, 651ft long, and lined in concrete.

The bridge consists of eleven fixed spans of 148ft, two end spans of 144ft each and a centrally pivoted opening span of 215ft. The spans consist of mild steel Pratt-type latticed girders supported on cast-iron cylinders, varying in diameter from 8ft to 12ft, infilled with concrete and founded on the underlying rock. Cross girders spanning between the latticed girders support the single line track. There is also overhead cross-bracing between the main girders. The contractor for the bridge was Sir William Arrol & Co. of Glasgow, the general contractor for the line between Rosslare and Waterford North Station being Sir Robert McAlpine & Sons. The route is currently closed to passenger services (HEW 3075)

L22. Nore Rail Viaduct (Thomastown) (S 576410). With the completion in 1850 of a viaduct over the River Nore to the south of Thomastown in County Kilkenny, the Waterford & Kilkenny Railway Company was able to push its line southwards to Waterford. The main span of the original viaduct was formed of timber-latticed girders and replaced in 1877 by the present structure, designed by C.R. Galwey and supplied by Courtney, Stephens and Bailey of Dublin. It consists of twin wrought-iron bowstring girders of 212ft span, bearing on cast-iron plates bedded into the rough-hewn coursed masonry abutments.

Nore Rail Viaduct, Thomastown

The depth of the girders at mid-span is 25ft 6in and 5ft 9in at the ends. The web of each girder consists of a diagonal bracing arranged in a double system of triangulation, dividing the flanges into fifteen equal bays, with stiffening plates at each end. Cross girders are suspended from the bottom flanges at the intersection of the diagonals. Each end of the main span is approached by twin masonry arch viaducts, the central piers of which incorporate transverse relieving arches. The total length of the viaduct is 428ft and the rail level is 78ft above riverbed level. The viaduct was refurbished in 2005 and a 'silent running' deck fitted. (HEW 3109)

Shannonbridge WATERWAYS IRELAND

L23. Shannonbridge (M 968255). As part of the improvement of the Shannon Navigation (**C1/C2**) in the 1840s, all the major bridges crossing the river were either replaced or altered by the inclusion of a navigation opening span. At Shannonbridge it was decided to retain the eighteenth-century multi-span masonry arch bridge, but to underpin the foundations and provide a twin-leaf cast-iron swivel bridge over the navigation channel at the eastern end of the crossing. The opening span was designed by Thomas Rhodes, chief engineer to the Shannon Commissioners, and manufactured and erected in 1843 by the Dublin firm of J. & R. Mallet. The swivel bridge was replaced in the 1980s by the present fixed beam and slab arrangement spanning the 65ft opening, the original bridge castings being preserved on plinths in an adjacent car park.

The bridge has an overall length of 692ft and contains a total of sixteen spans (excluding the navigation span), fourteen of 20ft span, and two end spans of 15ft. The arches are all semicircular and span between 6ft-thick piers. The bridge is only 16ft wide between parapets and traffic is single lane and controlled by lights. There is a fine example of a bridge-master's house at the eastern end of the bridge, designed by Thomas Omer around 1760 as part of the earlier improvements to the navigation. There are extensive fortifications at the western end of the bridge. (HEW 3132)

L24. Athlone Road Bridge, Weir and Locks (N 039416). The first masonry bridge at Athlone was erected around 1210. It was rebuilt in 1275 and replaced by a ferry in 1306. Sir Henry Sidney commissioned a nine or ten-arch masonry bridge, which was completed in 1567, the engineer being a Peter Lewys. A system of tolls operated up until 1826. Athlone Bridge at this time was only 14ft in width and the Shannon Commissioners' engineer, Thomas Rhodes, recommended its replacement by the present masonry arch bridge, each arch being of 61ft span and elliptical in form. This was built between 1841 and 1844 by the contractor John MacMahon. A 40ft cast-iron swivelling span at the western end was replaced in 1962 with a fixed reinforced concrete slab. Through traffic on the national primary route is now carried across the river on the Shannon Bridge, opened in 1991 as part of the Athlone bypass.

Immediately south of the town bridge are the large weir and locks of the Shannon Navigation (**C1/C2**). The old bypass canal to the west was abandoned in the 1840s and a new channel excavated in the riverbed. A new lock was built in the dry inside a large caisson. Caisson work was also necessary for the construction of the bridge piers. Major difficulties arose when constructing the weir, which became undermined. To allow work to proceed, the river was diverted down the old canal and the weir was then rebuilt. (HEW 3180)

Athlone Road Bridge *c.*1890 with opening span by Robert Mallet NATIONAL LIBRARY OF IRELAND

L25. Birr and Kinnity Suspension Bridges (N 053048 and N 203055). Birr Castle, beside the town of Birr in County Offaly, is the seat of the Earls of Rosse and the home of the Parsons family. The Camcor River, a tributary of the Little Brosna River, flows through the grounds of the castle. Some time before 1826, William Parsons commissioned the erection of a small iron suspension bridge of 44ft span across the Camcor. Thomas Cooke, writing in 1826, describes the bridge as ' a curious wire bridge'. Exact dating remains a problem but, following a fire in 1823, the castle was enlarged and this may well have been the time when the bridge was erected, although an earlier date is possible. The 34in-wide timber deck is carried on 4in by 2in timber stringers, which in turn are suspended by vertical wire hangers from a pair of cables. The suspension cables consist of eight ¼in-diameter forged iron wires, and are fixed at each end of the bridge to portal frames at a height of 5ft 6in above deck level. The portal frames are tied back to ground anchorages in the masonry abutments. The bridge has forged iron-latticed side panelling, typical of such bridges of the period. The portal frames were subsequently encased in concrete, probably around 1911, and this precludes an examination of the underlying original material. However, a similar-style bridge is located at Kinnity Castle, spanning the Camcor about 12 miles upstream. This has not been modified and appears to be identical in design to the

Birr Castle Suspension Bridge

bridge at Birr, with the exception of the span, which is only 26ft 6in. The portals are formed from pairs of 4in-diameter cast-iron tubing, connected across the top and suitably braced. The name of the maker stamped on the Kinnity portals is T. & D. Roberts, who operated a foundry in nearby Mountmellick. (HEW 3242)

L26. Obelisk Bridge (O 047758) was erected in 1868 to replace a timber bridge swept away by floods during the previous year. The bridge spans the River Boyne about 2 miles upstream from Drogheda at a point near to where an obelisk was erected to commemorate the Battle of the Boyne. It provides access to the Oldbridge estate and the Battle of the Boyne visitors' centre.

The bridge is made up of two wrought-iron double-latticed girders, each 128ft in length, bearing on four 4in-diameter 2ft-long expansion rollers, which in turn rest on bearing plates on the masonry abutments. The depth between the flanges of the girders is 10ft 8in, or one twelfth of the span. The existing abutment on the south side was reused, but the northern abutment was extended 28ft into the river in order to reduce the clear span to 120ft. The width between the top flanges of the girders is 16ft. The roadway is supported by buckled plates carried on shallow cross girders at 3ft 6in centres spanning between the

Obelisk Bridge

main girders. The prefabricated girders, each weighing over 28 tons, were transported from the works of Thomas Grendon & Co. in Drogheda upriver to the site on a pair of pontoons, and positioned on the abutments at high tide. Responsibility for the design of Obelisk Bridge was shared by the county surveyors of Louth and Meath (John Neville and Samuel Searanke) and the firm of civil engineers A. Tate. (HEW 3010)

L27. Vartry Reservoirs (O 218018 and O 200042). The first public water supplies for Dublin were drawn from wells, local streams and rivers, and later the canals. In 1860, the Hawkshaw Commission recommended that a gravity supply be provided from a surface reservoir to be constructed near Roundwood in County Wicklow, as originally proposed some years earlier by the engineering consultant Richard Hassard. The Dublin City Engineer, Parke Neville, drew up the plans for the scheme, Thomas Hawksley acting as consultant. The necessary Bill was steered through parliament in 1861 by Sir John Gray, Chairman of the Waterworks Committee, and the foundation stone laid at Stillorgan **(D24)** (the site of the Dublin service reservoirs) in November 1862.

The Lough Vartry impounding dam and reservoir are at an elevation of about 690ft above sea level and about 8 miles from the source of the River Vartry, which rises in the Wicklow Mountains to the west. The dam, consisting of a 1,640ft-long earthen embankment with puddle-clay core, reaches a height of 71ft. It is 27ft 10in wide at the crest and 377ft 3in wide at its lowest point. It carries the road from Roundwood to Wicklow along its crest. The 410-acre impounded reservoir has a capacity of around 2,500 million gallons and a greatest depth of about 60ft. The draw-off tunnel under the embankment, which accommodates the supply pipes from the draw-off tower to the treatment works, is 14ft square with a semicircular arched roof. Access to the intake tower is by a 97ft-span iron girder bridge consisting of twin latticed trusses with parabolic curved top chords. The main contractors for the 1862–67 scheme were William McCormick (London and Derry), all the ironwork being supplied and installed by Edington & Son of Glasgow.

A further reservoir to the north of Lough Vartry was constructed between 1908 and 1925 to the designs of John George O'Sullivan. The impounding earth dam of the upper

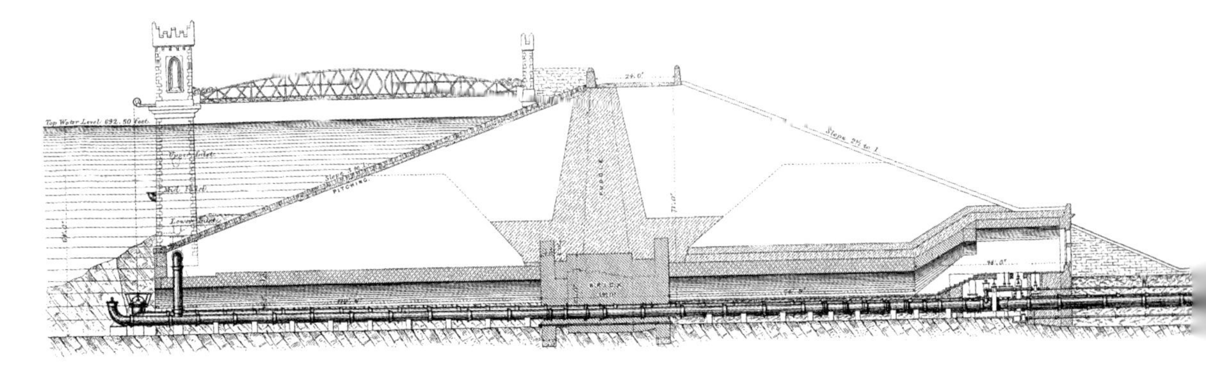

Section through Lough Vartry Dam near Roundwood

reservoir is somewhat longer and lower than the earlier dam. The reservoir covers an area of about 310 acres and has a capacity of around 1,200 million gallons. The main features are the two large section tunnels under the embankment (29ft wide by 26ft 3in and 23 ft 3in high respectively), and the bell-mouthed overflow shaft, 39ft deep with a circular ashlar granite weir 72ft in diameter. The larger tunnel connects the upper and lower reservoirs, whilst the other accommodates the overflow. The reservoirs are linked by a short steep-sided canal. The contractors were Kinlen, McCabe and McNally (1908) and H.S. Martin (Dublin) (1919). (HEW 3019 and 3020)

L28. Bohernabreena Reservoirs (O 092220). In 1883, the Commissioners for the Rathmines and Rathgar townships in County Dublin proceeded with a scheme to abstract water from the River Dodder above Bohernabreena in the Dublin Mountains, using the separation system. Two impounding reservoirs in the Glenasmole Valley were formed by the construction of earth dams. The upper dam is about 600ft long and reaches a maximum height above ground level of 70ft. The lower dam is of similar length and about 55ft high. The upper part of the catchment is an area of granite overlain by peat deposits. The resulting run-off is highly coloured and is intercepted and led around the upper reservoir in a bypass canal to the lower reservoir. This part of the supply was used originally as compensation water for the many mills along the lower reaches of the river, but the total supply is now piped to treatment works at Ballyboden. The larger upper reservoir receives its run-off from the remainder of the catchment area, which yields a purer water supply. The reservoir capacities are 360 million gallons and 160 million gallons respectively. The design, by Richard Hassard, was based on a scheme put forward by Parke Neville in 1854. The contractors were Falkiner and Stanford and the reservoirs were in operation by 1887. (HEW 3210)

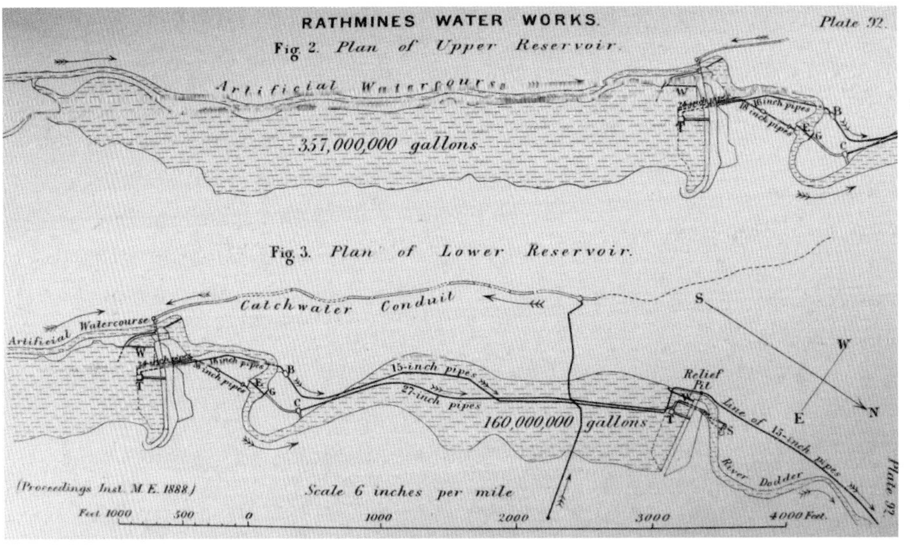

Reservoirs for Rathmines water supply (1888)

BOARD OF TRINITY COLLEGE DUBLIN

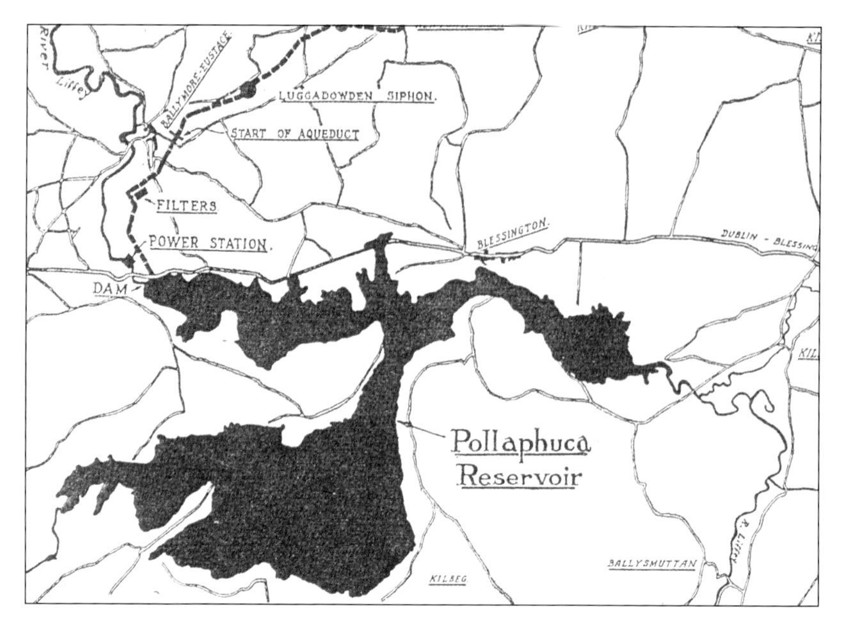

Dublin Water Supply (Blessington)

L29. Dublin Water Supply (Blessington). (N 945085) The growth of the Dublin suburbs in the 1930s resulted in increased water demand and the River Liffey was identified as the source for the additional supply. When it was learnt that the Electricity Supply Board (ESB) was planning a hydroelectric scheme on the river, Dublin Corporation agreed to become partners with the ESB in the construction of a dam at Pollaphuca in County Wicklow. Work on the scheme was carried out (with some difficulties occasioned by war conditions) between 1938 and 1943. The chief power plant and dam are at Pollaphuca, while a subsidiary power plant and dam were built 1½ miles downstream at Golden Falls. Together, these power plants have an installed capacity of 34 MW. Three new reinforced concrete bridges of portal frame design, with free spans suspended on cantilevers, were erected to carry roads diverted on account of the creation of the reservoir. The Pollaphuca Dam is gravity-type mass concrete, with a maximum height of 104ft and a total length of 225ft. There is a 1,230ft-long pressure gallery of 16ft internal diameter, a 900ft length being tunnelled in solid rock. Water is drawn from the reservoir through high-level and low-level tunnels, 390ft upstream of the dam, which lead to treatment works below the dam at Ballymore Eustace. From there the treated water enters a cast in situ ovoid-shaped concrete aqueduct 12.3 miles in length. This terminates near Rathcoole in south County Dublin, from where steel pipelines deliver the water to service reservoirs at Saggart, Belgard and Cookstown on the western fringes of the urban area. The aqueduct was constructed by the Cementation Company Ltd of Doncaster to designs prepared by the Corporation engineers, under the City Engineer, Norman Chance. The maximum internal dimensions of the aqueduct are 4ft 6½in high by 3ft 9½in wide. A second main was laid in 1953 and a new aqueduct, 25 miles long, completed to Fairview in the north of the city in 1994. (HEW 3259)

Trim water tower

L30. Trim Water Tower (N 817558) in County Meath is probably the oldest reinforced concrete water tower remaining intact in Ireland, having been erected in 1909. It was designed for the then Urban District Council by Mouchel & Partners of London, the contractor being the Irish Hennebique licensee, J. & R. Thompson of Belfast.

The cylindrical tank has an internal diameter of 30ft 6in and a height of 12ft. The wall thickness is 14in. The base of the tank is supported on a grillage of beams of four different sizes, carried on six legs. The 15in-square legs each have a 7ft-square pad foundation, all six pads being interconnected by a 'wheel' of ground beams. At 10ft vertical intervals, the 30ft-long legs are braced by a ring beam 12in by 9in in section. Inside, the legs are cross-braced in a triangle, a different set of legs being so braced at each level. The tower is now out of service, but remains in reasonably good structural order, given its age. (HEW 3257)

L31. Straffan Suspension Bridge (N 918295). Large country estates frequently included linear water features that necessitated the provision of at least one bridge. The landed gentry were able to afford iron or high-quality masonry arch bridges or, in some cases, chose suspension bridges, as at Birr Castle and as here at Straffan, the former estate of the Barton family in County Kildare, and now home to the K Club hotel and golf course. Spanning an artificial channel, fed from the River Liffey, this small footbridge was erected in 1849 by the firm of Courtney & Stephens of Dublin. The suspension bridge spans 45ft and is comprised of twin cast-iron portals carrying the suspension cables, the decking being suspended by vertical hangers from the cables. On top of the

Straffan Suspension Bridge

columns, there are ornamental finials. There is a handrail parallel to the decking, cross-braced with diagonal tie bars. The decking was later concreted in order to stiffen the structure for public use. (HEW 3038)

L32. Courtown Harbour (T 202563) lies on the County Wexford coastline east of Gorey. The prime promoter of the harbour here was Lord Courtown, who in 1824 obtained an Act of Parliament granting the necessary finance to the Courtown Harbour Commissioners to enable design and construction to begin. The contractors O'Hara, Simpson & McGill commenced work on a pier to the south at Breanoge Point and followed a northeastly direction as previously suggested in 1819 by the government engineer Alexander Nimmo. Asked for his opinion in 1825, George Halpin senior, the engineer to the Ballast Board at Dublin, foresaw problems of siltation. In 1831, Sir John Rennie's opinion was sought and by 1833 Jacob Owen of the Board of Works, in whose charge the harbour had been placed, reported that the pier was breaking up. The decision was then taken to build a new south pier at right angles to the shore near the mouth of the Breanoge River and to develop the harbour to more or less its present form. Under the direction of Francis Giles, in 1834–47,

Courtown Harbour

the south pier and then a new north pier were completed, the harbour was deepened and quays built. Sluice gates and an overfall were provided, and a system was designed to maintain the water level in the harbour at low tide and so provide a means of scouring the narrow entrance channel between the piers. The harbour was substantially completed by 1847. The south pier was rebuilt and lengthened in 1871, when Isaac Mann carried out repairs to the harbour. New sluice/tidal gates were provided in 1891 and in 1905 the responsibility for the harbour was transferred to Wexford County council, who carried out a partial rebuilding and lengthening of both piers in the 1960s. (HEW 3037)

L33. Wexford Harbour Improvement (T 065240 to T 095240 and T 060184 to T 079171). In 1846, the Wexford Harbour Improvement Company obtained an Act for constructing five embankments, with a view to enclosing much of the area of mudbanks (or slobs) in the harbour. The main objective was to confine the River Slaney to a narrower channel, so that increased scour, assisted by dredging, might create a permanent (south) deep-water approach to the quays at Wexford. The chief promoter of the scheme was J.E. Redmond, MP, assisted by William Dargan and others. The company appointed James B. Farrell to be its engineer.

Work commenced on enclosing the north slobs by constructing a clay- and marl-impounding embankment from the public road at Ardcavan, 4,800ft in a straight line to

Pumping Station at North Slobs, Wexford Harbour

Big Island, thence curving slightly to the south a further 5,490ft to Raven Point. The crest of this embankment is 6ft wide and rises 13ft 6in above low water ordinary spring tides. The seaward slope is pitched with stone at 1 in 1½, the landward slope being 1 in 1. An 18ft-wide roadway and a 4ft-wide drain were provided on the landward side. A pumping station was erected south of Ardcavan church, power to the pumps being supplied first by a steam engine, and latterly by electricity. Water is pumped from the Bergerin and Curracloe catchment drains into the sea at low tide through sluices in the embankment.

After completion of the north slobs enclosure works, a further Act was obtained in 1852 to carry out similar, but less extensive, works to enclose slobs on the south side of the harbour, a pumping station being erected at Drinagh. The Drinagh station is now derelict, but the pumping station building at the north slobs was restored by the Office of Public Works. Since 1969, parts of the north slobs have been occupied by the Wexford Wildfowl Reserve and are a wintering ground for about half the European population of Greenland white-fronted geese. (HEW 3067)

Rosslare Europort IARNRÓD ÉIREANN

L34. Rosslare Harbour (T 136126) is located at the southeast corner of County Wexford near Greenore Point, and about 8 miles southeast of Wexford town. In 1873, the Rosslare Harbour Commissioners built a 480ft-long timber jetty to provide two berths for steamers from Fishguard in South Wales. This was connected to the shore by a 1,000ft-long viaduct. When the Dublin, Wicklow & Wexford Railway was extended from Wexford to Rosslare in 1882, a new jetty and viaduct were provided, the jetty being dog-legged and mostly of iron and timber construction.

Following the setting up of the Fishguard & Rosslare Railways & Harbours Company jointly by the Great Western Railway in England and the Great Southern & Western Railway, a new harbour was built between 1904 and 1906. This consisted of a double-track railway viaduct of eleven 81ft spans of steel girders on cast-iron cylindrical piers. This led to a new jetty, 1,550ft in length. New quay and sea protection walls were also completed in mass concrete and land reclaimed from the sea using sand obtained from suction dredging of the approaches to the harbour. Road access was provided by a reinforced concrete bowstring girder bridge spanning the rail tracks. The main contractor for the 1882 works was W. Murphy of Dublin and for the 1906 contract, Charles Brand & Company of Glasgow. Only traces of the works described above now remain as the harbour was completely rebuilt and enlarged, first in the late 1960s and more recently in the 1990s. The harbour is a major ferry terminal and container port, now renamed Rosslare Europort. (HEW 3092)

L35. Hook Head Lighthouse (X 732972). Generally regarded as the site of one of the oldest lighthouses in either Britain or Ireland, the present lighthouse on a headland in the southwest of County Wexford began as a fortress and lookout tower. This was built by the Norman, Raymond Le Gros, in 1172 at the entrance to Waterford Harbour. Welsh monks are reputed to have maintained a crude form of warning beacon here from as early as the fifth century. In the thirteenth century, William Marshall established a light on the tower to guide ships up the estuary to the new port of Ross (now New Ross). The light was an open fire maintained by local monks from nearby Churchtown.

The tower was altered and enlarged over the years, the present lantern and supporting turret being erected in 1864 to replace a smaller lantern and turret of 1792. The tower has three vaulted chambers and a spiral staircase. The walls of the tower vary in thickness from 12ft 10in to 9ft. The earlier part of the tower is about 75ft in height (four storeys) with a diameter of about 38ft. The final 20ft of the tower to the level of the balcony is about 20ft in diameter. The overall height to the top of the lantern is 118ft. The diameter of the main tower is about 40ft, but narrows somewhat towards the top. Lighting was by oil lamps from 1791 until 1871, when coal gas, provided from a small gasworks, was used. This was replaced by vaporised paraffin in 1911 and then by electricity in 1972. The

Hook Head Lighthouse

families of lighthouse keepers lived at the Hook until 1977 when the lighthouse became 'relieving'. It was converted to automatic working in 1996. There is now a visitors' centre at the lighthouse. (HEW 3068)

L36. Kish Bank Lighthouse

(O 392304). At the eastern approach to Dublin Bay lies the Kish Bank, a hazard to shipping first marked in 1811 by a light vessel stationed at the northern end of the bank. In 1842, Alexander Mitchell attempted to erect a lighthouse on a platform supported by his patent screwed piles, but bad weather caused the attempt to be abandoned. The bank continued to be marked by a light vessel until the 1960s when the Commissioners of Irish Lights decided to build the present structure.

Kish Bank Lighthouse
COMMISSIONERS OF IRISH LIGHTS

The resulting lighthouse is a telescopic reinforced concrete structure, comprising an outer caisson of three concentric cylinders of varying height standing on a 3ft-thick base slab and interlocked by twelve radial walls. Outside the centre cylinder, which contains the tower, the caisson is sealed by roof slabs. The diameter of the outer caisson is 104ft, that of the inner caisson being 41ft. The total height of the tower from the basement to the helicopter platform is 112ft. The lower 13ft of the tower is 35ft 6in in diameter, that of the upper 57ft 6in being 17ft 6in. The tower was constructed inside a caisson at Traders Wharf in Dun Laoghaire Harbour, floated out to the site and sunk onto a prepared bed. The telescopic tower was then floated up to near full height by allowing water to enter the caisson. The water was then pumped out, and the space filled with sand and sealed with concrete. The lighthouse became operational on 9 November 1965 and was provided with the usual accommodation and operational facilities. The light is now operated automatically and is only visited occasionally for maintenance purposes. The Kish Bank lighthouse was designed by R. Gellerstad and built by Christiani and Nielsen of London and Copenhagen. (HEW 3050).

L37. St John's Bridge, Kilkenny (S 508558). When completed in 1912, St John's Bridge was the longest-span reinforced concrete road bridge in either Britain or Ireland. Crossing the River Nore in Kilkenny, this reinforced concrete arch of 140ft clear span is an early example of the use of what was then popularly known as 'ferroconcrete'. It replaced a three-span masonry arch bridge erected in the 1760s.

The design of the present bridge was by Mouchel & Partners of London and Alexander Burden, the county surveyor of Kilkenny. The work was carried out by J. & R. Thompson of Belfast and Dublin, a licensee of the Hennibique system of reinforcing concrete. The bridge has three arched ribs 2ft wide and spaced 12ft apart centre-to-centre, spanning from concrete buttresses, which are carried on precast piled foundations. Each rib is 2ft 6in deep at the springings and 2ft deep at the crown. The main longitudinal beams supported by the arched ribs are of 14in by 24in section. When built, the bridge had open spandrels consisting of fourteen columns of 12in-square section. In 1970, it was found that the piles had moved horizontally and it was decided to remove the cobbled road surfacing and place a concrete slab over the deck. At the same time, the spandrels were filled in, thus radically altering the way the structure reacts to the loads placed upon it. (HEW 3167)

St John's Bridge, Kilkenny

L38. 'The Deeps' Bridge, Killurin (S 975268). The River Slaney between Enniscorthy and Ferrycarrig in County Wexford is spanned by an early reinforced concrete road bridge at an area known as 'The Deeps' near Killurin. The present bridge replaced a timber trestle bridge erected in 1842–44. The bridge consists, on the eastern side of the waterway, of five spans, each of 30ft, approached by a 216ft-long embankment across the flood plain terminating in a masonry abutment. On the western side there are five spans, four of 30ft and one of 19ft 6in. Between these spans there is a bascule-type steel lifting span of 40ft navigation opening, placed between two fixed sections 10ft 6in long. The supporting piers consist of reinforced concrete pile groups carried up to the deck and cross-braced. The overall length of the bridge is 609ft, 362ft being over the waterway. The overall width is 20ft 4in and 18ft 4in between parapets. The bridge was designed by Delap and Waller of Dublin and erected in 1915 by the British Reinforced Concrete Engineering Company, using their own patented method of reinforcement. The lifting span was supplied by the Cleveland Engineering Company. (HEW 3179)

'The Deeps' Bridge, Killurin, County Wexford

Five-sailer windmill at Skerries, County Dublin

L39. Skerries Windmills (O 250605). The larger of two windmills at Skerries in north County Dublin (Fingal) forms part of a water- and wind-power complex located in the centre of a town park. Known as 'The Great Windmill of Skerries', the present four-and-a-half-storey tower mill is a reconstruction of the mill that suffered a disastrous fire during a gale about 1860. The base of the tower is 20ft in diameter, reducing to 17ft 1in at the base of the cap. The walls reduce in thickness from 2ft 8in thick at the base to 2ft 1in at cap level about 34ft above ground level. The overall height to the top of boat-shaped cap is 49ft. The weight of the cap, including the sails, is estimated to be around 17 tons. There are five sails, each 33ft in length, the system of sails being of a primitive spring-type consisting of twenty louvres to each arm, connected together, spring loaded and adjusted by hand. The cap has a tail pole and is manually rotated into the wind. The wind shaft is inclined at 16° to the horizontal. The mill ceased operation as a corn mill around the beginning of the nineteenth century and was restored in 1995–97. There is also a smaller four-sail windmill on site and a working waterwheel. (HEW 3264)

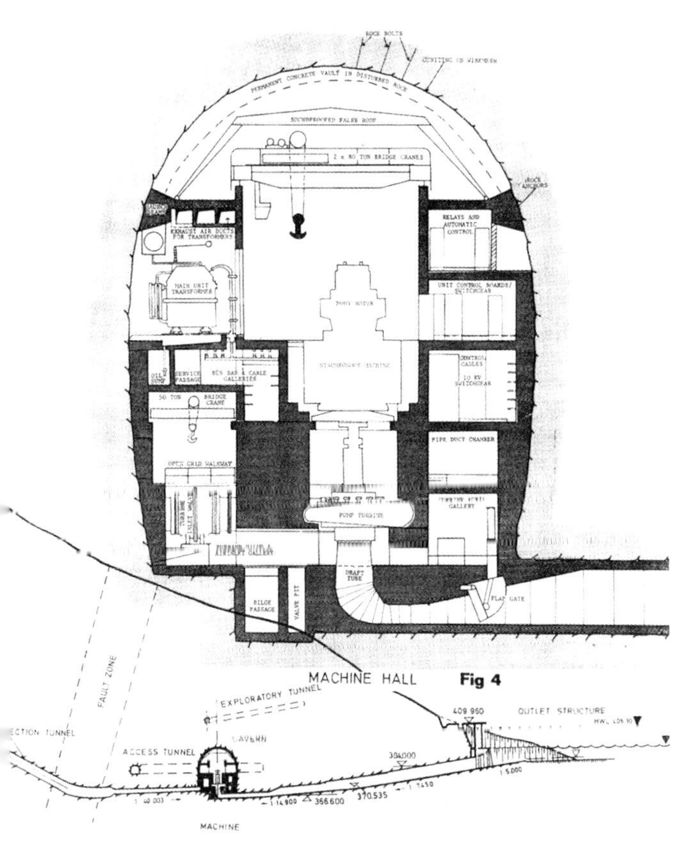

Turlough Hill Pumped Storage: Machine Hall
ESB ARCHIVES

L40. Turlough Hill Pumped Storage Station (T 090990). Ireland's only pumped storage electricity generating station is situated at Turlough Hill near Glendalough in County Wicklow. It is not a primary producer of electricity, but uses the excess capacity that is available in the national system during periods of low demand – mainly at night – to pump water from a lower reservoir to an upper reservoir. It then uses this same water to produce electricity during periods of high demand. It can also be used to provide electricity on demand at times of sudden loss of system generating capacity.

The scheme was designed by the Civil Works Department of the Electricity Supply Board. The cavern and tunnels were constructed between 1969 and 1974 by a consortium of German companies led by Alfred Kunz & Co. in association with the Irish Engineering and Harbour Construction Co. The station was commissioned in 1974. The upper artificial reservoir, which holds 2.3 million cubic metres of water, was formed on the top of the mountain by removing many tonnes of peat overburden. Rock fill was then excavated and used to form a 25m-high embankment, and the inside was lined with asphalt and sealed to make it watertight. The lower reservoir was a natural lake, Lough Nahanagan, the bed of which was lowered by about 15m. The mean gross head available at the site is 287m. The cavern inside the mountain is 82m long by 23m wide by 32m high and entailed the removal of 47,000 cubic metres of granite rock. The cavern houses four 73 MW reversible pump turbines and associated generating units. A single steel-lined pressure shaft connects the turbines with the upper reservoir and has an internal diameter of 4.8m and a length of 584m at a slope of 28°. The tailrace tunnel has a diameter of 7.2m and is 106m long, whilst the main access tunnel to the cavern is 565m long and has a curved segmental section of maximum height 5.5m above a flat floor 5.7m wide. (HEW 3226)

DUBLIN CITY & DISTRICT

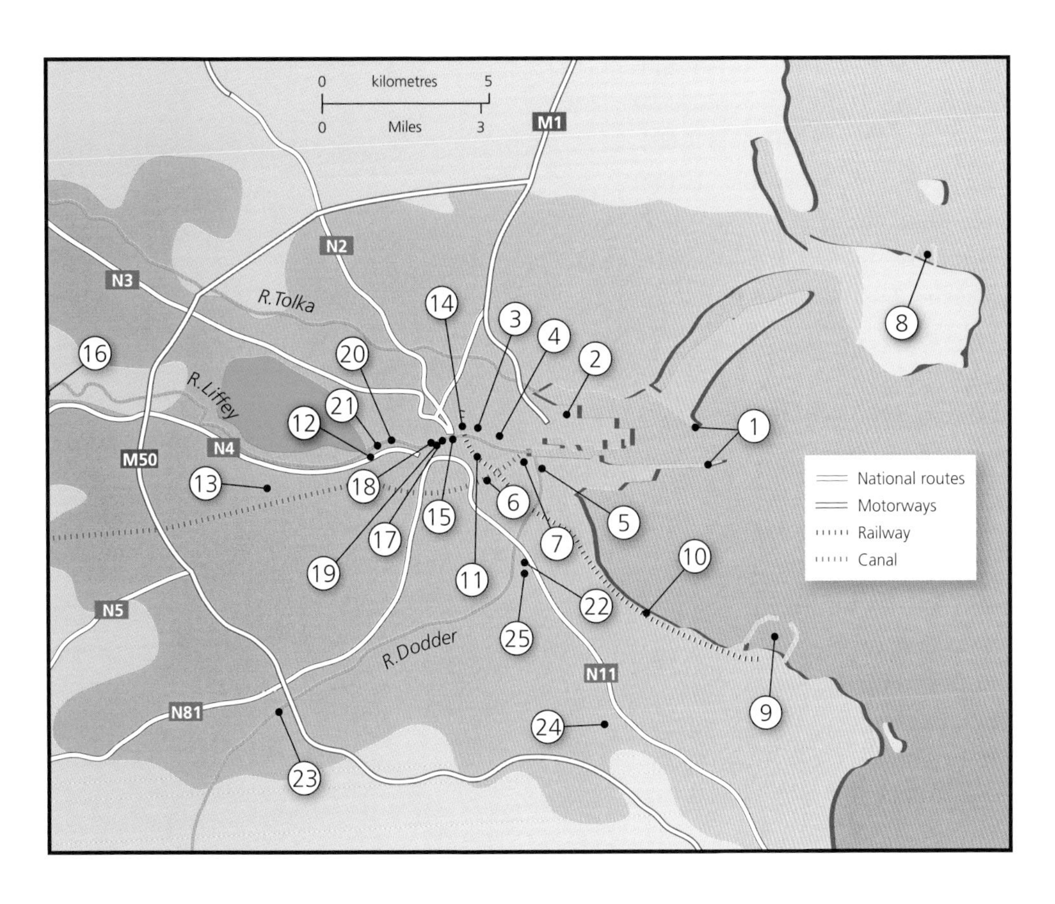

D1 The Bull Walls
D2 Dublin Port
D3 Custom House Docks and Stack 'A' Warehouse
D4 Scherzer Lift Bridges, Dublin
D5 Ringsend Bridge
D6 Grand Canal (Circular Line)
D7 Grand Canal Dock
D8 Howth Harbour
D9 Dun Laoghaire Harbour
D10 Dublin & Kingstown Railway
D11 Westland Row (Pearse) Train Shed
D12 Heuston Station and Train Shed
D13 Inchicore Railway Works
D14 Liffey Rail Viaduct
D15 Butt Bridge
D16 Lucan Bridge
D17 O'Connell Bridge
D18 Grattan Bridge
D19 Liffey Bridge
D20 Rory O'More Bridge
D21 Sean Heuston Bridge
D22 Donnybrook Bus Garage
D23 Balrothery Weir
D24 Vartry Aqueduct and Stillorgan Reservoirs
D25 UCD Water Tower

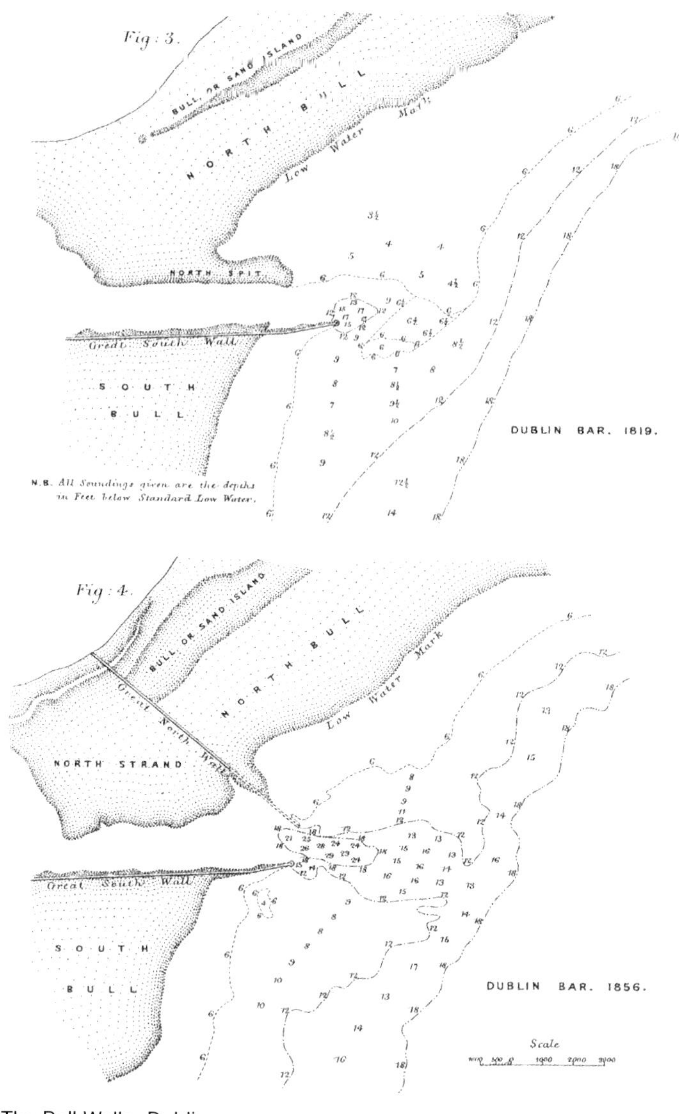

The Bull Walls, Dublin

D1. The Bull Walls.

(O 181342 to O 232340 and O 211360 to O 232344) The city and port of Dublin are situated where the rivers Liffey, Dodder and Tolka discharge into Dublin Bay. Large accumulations of sea sand to the north and south of the river estuaries form what are known respectively as the North and South Bulls. At the beginning of the eighteenth century, the rivers flowed out over these strands at low water and formed natural, but constantly shifting, channels.

In 1711, work began on the provision of a straight channel from the city to Ringsend and a jetty extending 7,938ft from Ringsend to the site of the Pigeon House Fort, and thence a distance of 9,816ft to the eastern spit of the South Bull. This jetty was formed of timber gabions. These were constructed as caissons at Ringsend and floated down river to the site, where they were then filled with rubble to sink them to the seabed. The jetty was completed in the 1730s and a floating light vessel moored at its extremity. The first section, from Ringsend to the Pigeon House Fort, was replaced in 1748 by a double line of rubble retaining walls with sand infill. The Poolbeg Lighthouse was completed at the end of the timber jetty in 1767. The present South Bull Wall was completed in 1796 by the Corporation for Preserving and Improving the Port of Dublin (commonly known as the Ballast Board), who also provided a small harbour at the Pigeon House to shelter vessels from easterly winds.

About 1800, the Board commissioned Captain William Bligh (of *Bounty* fame) and others to survey the harbour and report on what improvements could be made, particularly with regard to increasing the depth of water at the entrance to the port. The final solution decided on was a scheme, proposed by Francis Giles and George Halpin senior in 1819,

Alexandra Basin, Dublin, early twentieth century DUBLIN PORT ARCHIVES

which involved the construction of a wall or breakwater from the north shore at Clontarf to a point opposite the Poolbeg Lighthouse. Completed in 1824, the North Bull Wall is formed of rubble limestone and granite with an opening between the Bull Island and the shore spanned by a wooden bridge. From the end of the bridge, a length of 5,500ft of wall was completed to the full design height, a length of 1,500ft to high-water neap tides, and the remaining section of 500ft to half-tide level. On the recommendation of Thomas Telford in 1822, a further length of 500ft was added and the following year, a further 1,000ft, bringing the total length to 9,000ft. On the north side of the bay, during the first half of its ebb, the tide runs westwards towards the bar, and thence in the direction of Dun Laoghaire, whilst during the last half of its ebb and the whole of its flood, the tide sets eastwards towards the Baily Lighthouse at the extremity of the Howth peninsula. During the first half of the ebb tide, the tidal and river waters, confined within the North and South Bull walls, pass partly over the submerged portion of the North Bull Wall, and partly through the harbour entrance between the walls. As soon as the ebbing tide uncovers the lower portion of the wall, the remainder of the tidal and river waters within the harbour are confined to pass through the narrow entrance. The result is a great increase in the velocity of the current and the removal by scouring action of sand from the bar at a time when the ebb tide is setting eastwards towards deep water. Due to the direction of the currents on the north side of the bay, the flood tide from the south does not normally return the sand to the harbour. As a result of this remarkable piece of harbour engineering, the depth at low-water ordinary spring tides of the navigable channel through the bar had by 1873 increased from around 6ft to over 16ft. (HEW 3016)

Custom House Docks, Dublin, prior to redevelopment DUBLIN PORT COMPANY

D3. Custom House Docks and Stack A Warehouse (CHQ) (O 166346). A decision by the port authorities towards the end of the eighteenth century to provide deep-water berthage further east of the Custom House resulted in the construction of the Custom House Docks. George's Dock (320ft by 250ft), with a separate tidal lock entrance, was opened in 1821. An Inner Dock (556ft by 279ft) was completed by 1824. Both docks, together with the large warehouses, were designed by John Rennie and completed by Thomas Telford following Rennie's death in 1821. Later the docks were developed mainly for warehousing. When the beginnings of the deep-water port were created further down river, the docks fell into disuse, as shipping was able to enter the port at all states of the tide, instead of having to enter the wet docks at high tide through a system of locks. However, the warehouses continued to be used for storing goods landed in other sections of the port. Stack A Warehouse was designed by Rennie and erected in the early 1820s as a tobacco warehouse. There are four 154ft wide bays in a total length of 476ft. The warehouse was originally 500ft long but was later shortened to permit widening of the quayside roadway. The clear internal height is 20ft, but a mezzanine floor 154ft long was added at the north end in 1871. The roof trusses span 38ft 6in and are supported, in the case of the two outer bays, on 2ft 3in-thick masonry walls and, internally, at the quarter points by independent cast-iron arches spanning 18ft between 9in-diameter cast-iron columns. These in turn bear on the masonry walls of the vaults underneath. The trusses are formed from cast-iron compression members of cruciform section and wrought-iron tension members of circular section. Each truss is supported by pairs of struts extending at 45° from the column heads to points on the truss approximately 3ft from each side. The tops of the inclined struts are tied together by a circular-section tie bar, thus completing the triangle of forces. The curved longitudinal beams on top of the columns are 36in at the springing, have hollow spandrels, and a rise-to-span ratio of about 0.3. The iron roof purlins are T-shaped with tapered flanges and webs. In 2007 the warehouse was rehabilitated and renamed CHQ (Custom House Quay) to accommodate a range of commercial activities, the only significant change being the replacement of the southern plain brick wall facing the river with a planar glass wall set back to allow light into the vaults. (HEW 3053 and 3066)

D4. Scherzer Lift Bridges, Dublin (O 173345). The Royal Canal joins the River Liffey in Dublin at the North Wall quays to the west of the East Link Bridge and is crossed at this point by twin lifting bridges. These bridges were erected in 1912 by Spencer & Co. of Melksham, Wiltshire to replace a rolling drawbridge placed there in 1860. Sir John Purser Griffith, chief engineer to the Dublin Port and Docks Board, based his design on that previously patented in 1893 by William Scherzer of Chicago.

Each bridge consists of two main girders, connected together by fourteen floor beams, the floor being composed of patented 'buckled plates' resting on and riveted to the floor beams and to joists fitted between the floor beams. To the western ends of the main girders,

Scherzer lifting bridges over the entrance to the Royal Canal in Dublin

segmental girders are attached to form the rolling surfaces upon which the bridges bear. The segmental girders are extended so as to carry a large iron and concrete counterweight and the whole structure is suitably braced. The centre of gravity of the whole moveable structure (now fixed in the open position) corresponds with the centre of roll, thus ensuring that the bridge was in equilibrium in all positions. To prevent accidental displacement of the bridges on their paths, teeth were formed on the cast-steel track plates, and corresponding recesses cut and accurately tooled in the curved track plates of the segmental girders. Each bridge was worked by electric motors (since removed), erected on the platform in front of the counterweight box, but they could be operated manually in case of power failure. A similar pair of bridges was erected by Sir William Arrol of Glasgow in 1932 over the entrance to the Custom House Docks further upriver. (HEW 3015)

D5. Ringsend Bridge (O 180337). The old road from Dublin to Ringsend crosses the River Dodder at Ringsend by a masonry bridge of an interesting design. The present bridge, designed and built by Dublin Corporation is a single-span masonry arch bridge constructed using dressed Wicklow granite and replaces an earlier bridge carried away in the flood of 1802. Built in 1812, the Ringsend Bridge spans 78ft between abutments and is 27ft wide between the solid parapets. The rise of the span is 14ft, and the arch has an elliptical profile. The edges of the arch are chamfered in the manner of the French *cornes de vache*, the chamfer reducing from the springing level to the top of the arch. The abutments are continued downwards as a curve to form a paved riverbed. Both these elements of the design ensure good hydraulic flow conditions under the bridge in times of heavy flooding. The spandrel walls are formed by a continuation of the voussoir stones in the arch ring. There is a string course running the full length of the bridge immediately underneath the parapet. (HEW 3047)

Ringsend Bridge, Dublin

D6. Grand Canal (Circular Line). (O 128332 to O 174336) The Dublin terminus of the main line of the Grand Canal (**L8**) connecting Dublin with the Shannon Navigation (**C1/C2**) was at James's Street Harbour near the present Guinness Brewery at a point where there is a steep drop down to the River Liffey. William Chapman surveyed a route for a canal to link the main line of the Grand Canal from the 1st Lock at Suir Road to the river near the mouth of the River Dodder at Ringsend.

The canal (known as the Circular Line) was constructed between 1790 and 1796 under the direction of William Rhodes, James Oates and (for a time) Archibald Millar, all engineers to the Grand Canal Company. The canal cost nearly £57,000, almost five times the original estimate. The canal is about 3½ miles long and is level from Suir Road as far as Portobello, where a harbour and a hotel were built in 1805. Portobello became the terminus for the passage boats plying the canal between Dublin and Limerick. From Portobello the canal falls by a series of seven single locks to a large dock built by the canal company near Ringsend. (HEW 3228)

Grand Canal (Circular Line)

Grand Canal Dock, Dublin

D7. Grand Canal Dock. (O 178339) A river wall, approximately 3,000ft long, was constructed between 1717 and 1727 from the end of the south quays to near the mouth of the River Dodder (now known as Sir John Rogerson's Quay). By 1760 a bank had also been built along the line of the present South Lotts Road. The area within the banks was reclaimed, together with adjoining areas of the Dodder estuary. Within this area, the Grand Canal Company built a large L-shaped wet dock, divided by the main road to Ringsend into two areas of 16½ and 8 acres respectively and providing 5,300ft of wharfage.

The Grand Canal Dock was constructed between 1792 and 1796, the work being supervised by Edward Chapman, a brother of William Chapman, who acted as consultant to the project along with William Jessop. Jessop had earlier drawn up plans for the dock and designed the entrance gates, but left much of the implementation of the project to Chapman. The contractor was John Macartney, who was knighted by the Lord Lieutenant at the official opening of the dock on 23 April 1796. Part of the dock to the west was later filled in after the railway was opened in 1834. The dock was never a commercial success, because of the small size of the locks, siltation problems in the river, and the decision by the Ballast Board to build new docks near the Custom House upstream and on the opposite bank of the River Liffey (**D3**). There are three parallel tidal locks of varying capacity: Camden Lock (awaiting restoration), Buckingham Lock and Westmoreland Lock (both restored). The canal company also built three graving docks on the land between the docks and the Dodder estuary, the largest being filled in as early as 1851. At this time the Dublin Dockyard Company took a lease on the area and continued in business until 1881. The Ringsend Dockyard Company operated in and around the two smaller graving docks from 1913 until 1963, when they too were filled in. The smallest of the graving docks is currently undergoing restoration. Much of the dock area has now been redeveloped for office, residential and leisure purposes. (HEW 3082)

D8. Howth Harbour (O 286395). An insufficient depth of water at Pigeon House Harbour and a general shortage of berths in Dublin port were causing such delays to the mails from London that, from 1800 onwards, varying proposals were put forward by Thomas Rogers, Sir Thomas Hyde Page, and John Rennie for an alternative harbour to be constructed at Howth on the north side of the peninsula to the east of Dublin. One proposal included a canal to connect the harbour with the city.

In 1807 a start was made on the construction of an east pier at Howth to a plan drawn up by Captain George Taylor, who initially supervised the work. In 1809, about 240ft of the end of the pier collapsed and Taylor resigned, being succeeded as superintendent of

Howth Harbour Lighthouse

the work by John Aird, acting under the direction of Rennie. Rennie had advocated Dun Laoghaire (**D9**) as a more suitable site for an asylum harbour, but work had already started at Howth. Following a recommendation in 1810 by Rennie, a west pier was added and the harbour was substantially completed by 1813. By 1812, Rennie had been able to use a diving bell to assist in the completion of the pier heads, but prior to this, pier foundations below water were prepared using Runcorn stone, the soft texture of the stone making it easy to cut into blocks, each four tons in weight. These blocks, when submerged in water, acquired a hard texture and were laid sideways on in front of each other at an angle of 40°. On top of these, horizontal header and stretcher courses were laid in mortar. Howth was not formally established as the mail packet station until 1818, when a lighthouse was added at the end of the east pier, together with a light keeper's dwelling (converted to two storeys in 1856). The original harbour was formed by two piers, each 200ft wide at the base and 85ft wide at the high-water mark, with a 42ft wide roadway on top. The east pier

extended 1,300ft in a north-northeasterly direction, then 230ft north-northwest and finally 920ft northwest by west. The west pier, commencing approximately 2,000ft along the shoreline, is 1,900ft in length. The harbour entrance is 350ft wide and the area enclosed is some 52 acres. To counteract wave action, 100ft-long moles have since been added at the extremities of the piers.

The harbour was a failure as far as the mail packet ships were concerned and the mails were transferred to Kingstown (Dun Laoghaire) in 1826. Howth Harbour became instead a major fishing port and yachting centre, with extensive marina facilities. (HEW 3055)

D9. Dun Laoghaire Harbour (O 245294). In 1815, eight harbour commissioners were appointed for the purpose of arranging for the building of a new harbour, eastwards of the old fishing port of Dunleary, to replace an earlier pier. This had rounded pier heads and a parapet, and was completed in 1767. The 'Dunleary Asylum Harbour' was intended primarily to provide a safe refuge for ships unable to reach Dublin during heavy winter gales. The initial design of the harbour consisted of a single pier to be carried out about 2,800ft from the shoreline, but in 1817, John Rennie was consulted and subsequently proposed two embracing piers, which later became known as the East and West Piers. In the same year the foundation stone for the harbour was laid by the Lord Lieutenant of Ireland, Earl Whitworth. Between 1817 and April 1820, 2,285ft of the East Pier was completed. Three years later it had reached 3,350ft, its final length being 4,231ft. The West Pier, begun in 1820, was already 1,600ft long by 1823, and by December 1827 had reached 4,140ft; its final length was 5,077ft. The base of each pier is 310ft wide and constructed with blocks of Runcorn sandstone, each 50cu.ft in volume. From 6ft below low water and upwards, granite was used. At the top, the pier is 52ft wide, with a 40ft promenade on the inner side and an 8–9ft-high parapet wall to protect the upper promenade from waves. The piers enclose an area of about 250 acres. The

Dun Laoghaire Harbour
DUN LAOGHAIRE HARBOUR COMPANY

rock for forming the piers was quarried at nearby Dalkey and transported to the harbour by means of a funicular railway using six trucks connected together by a continuous chain, each truck carrying 25 tons of granite. The core of the piers consists of granite rubble loosely tipped and allowed to consolidate using the action of the waves.

Following the death of John Rennie in 1821, his son John (later Sir John) Rennie, took over as consultant to the project. A decision on the final form and length of the pier heads was delayed until the Board of Works took over responsibility for the harbour in 1833, the final outcome being to form an entrance to the harbour of about 760ft in width, with rounded pier heads. The present East Pier Lighthouse and Battery Fort were completed in 1860.

Mail packet steamers, on the service to Holyhead, were transferred from Howth (**D8**) in 1826 and berthed at a wharf near the present bandstand on the East Pier. A 500ft wharf was completed in 1837, Traders Wharf in 1855 and the Carlisle Pier with direct rail connection in 1859. A car-ferry terminal was built in 1970 and the rail link finally abandoned in 1981. Responsibility for the harbour was transferred in 1989 from the Office of Public Works to the Department of the Marine & Natural Resources. The harbour is now under the control of the Dun Laoghaire Harbour Company, a state-owned commercial company. A major extension to the vehicle and passenger handling facilities was completed by ASCON Ltd in 1995, including a berth for a new high-speed catamaran passenger/vehicle ferry. This ferry was the largest of its type in the world at the time of its introduction in 1996 on the Holyhead–Dun Laoghaire service. Extensive marina facilities have also been developed in recent years. (HEW 3014)

D10. Dublin & Kingstown Railway (O 166340 to O 243288). An Act of Parliament in September 1831 gave royal assent to a scheme prepared by Alexander Nimmo for 'making and maintaining a railroad from Westland Row in the City of Dublin, to the head of the western pier of the Royal Harbour at Kingstown in the county of Dublin with branches to communicate therewith'. Nimmo died in January the following year and Charles Blacker Vignoles was appointed in April 1832 by the Board of Works to report on the scheme, subsequently becoming engineer to the Dublin & Kingstown Railway Company (D&KR). This, the first public railway line in Ireland, is double track and was laid initially to the English gauge of 4ft 8½in. The line was converted to the Irish gauge of 5ft 3in when the line was leased in 1854 by the Dublin and Wicklow Railway Company, which extended the railway to Bray.

From the original terminal station at Westland Row (now Pearse) Station (**D11**), the line runs on an embankment between retaining walls as far as Serpentine Avenue in Ballsbridge and thence to Merrion on a low earth embankment. From Merrion to Blackrock and from Seapoint to Dunleary (renamed Kingstown in 1828 to mark the visit of William IV), the

Dublin & Kingstown Railway (opened 1834)

line is carried on embankments laid on the strand. These sea embankments comprise two parallel bunds consisting of clay/gravel filled in between with sea sand, topped with layers of gravel, and followed by the track ballast. The slopes are pitched with granite blocks and topped with heavy parapet walls.

The initial length of the line was about 5½ miles, but it was extended in 1837 a further half mile to the mail steamer wharf in the Royal Harbour (**D9**). This cut off access to the old harbour at Dunleary and a compensation harbour was built (now known as the Coal Harbour). The embankment, boat slips and harbour arrangements were designed by William Cubitt and constructed, as was the entire project, by William Dargan. The present station buildings at Dun Laoghaire were designed by the architect John Skipton Mulvany, the walls of the temporary train shed erected in 1845 being seen today to the northeast of the present platforms. A single-track extension to Dalkey, built by Dargan in 1844, was worked on the atmospheric principle developed by Samuda and Clegg and acted as a feeder to the D&KR until April 1854. In 1884, parts of the original train shed at Westland Row were incorporated into the rebuilt terminus, and provision was made to allow for subsequent through working of trains across the River Liffey to Amiens Street (now Connolly Station) via the Liffey Viaduct or Loop Line (**D14**). The station layout at Dun Laoghaire was altered in 1856 to permit through working to Bray and was further modified in 1985 by the addition of a platform on the south side as part of the electrification of the line, representing the first phase of the Dublin Area Rapid Transit System or DART. (HEW 3025)

D11. Westland Row (Pearse) Train Shed (O 166340). Westland Row (now Pearse) Station was originally the terminus of the Dublin & Kingstown Railway (**D10**). The present train shed was completed in 1884 for the Dublin, Wicklow & Wexford Railway. The roof design is based on that first devised by the famous Dublin iron founder, Richard Turner, for Lime Street Station at Liverpool, the plans being obtained from his son William, also an iron founder. The principals and arched girders were made in Chepstow, but the rest of the ironwork was supplied by the Dublin firm of Courtney, Stephens & Bailey.

The roof of the main bay of the train shed spans 88ft 10in, the smaller bay to the southwest having a span of 64ft 8in. The top chords of each principal truss are on radii of 56ft 5in and 41ft 1½in respectively. The principals are at 13ft 4in centres and the columns at 40ft centres. The main roof is 510ft long, the smaller roof about 240ft. The principals rest on arched girders spanning longitudinally between the columns.

In 1890–91 the northwest end of the station was further modified to permit trains to connect with Amiens Street (now Connolly) station on the north bank of the river via the Liffey Viaduct. A bowstring girder with curved bottom and top chords was inserted to span the main through platforms and thus maintain the structural integrity of the roof. A new street-facing facade was built, consisting of wrought-iron pillars inserted between the existing masonry abutments and a facia formed of decorative cast-iron panels to match the ornamentation on a new bridge spanning Westland Row. (HEW 3222)

Westland Row (Pearse) train shed

Train shed at Heuston Station, Dublin

D12. Heuston Station and Train Shed (O 138343). The Great Southern & Western Railway (GS&WR) headquarters in Dublin was at Kingsbridge (now Heuston) Station. The station consists of an elaborate main building, designed by the English architect Sancton Wood, a pupil of Sir Robert Smirke. It was completed in 1844, and a large train shed to the rear, designed by Sir John Macneill, engineer to the GS&WR, was completed four years later. The main building consists of a central section, two storeys high, the lower part rusticated, from which spring Corinthian pillars supporting the cornice surmounting the upper storey. Clock towers rise from each of the single-storey flanking wings. The original train shed covers four platforms (Nos. 2 to 5) and is 616ft long and 162ft wide. Seventy-two cast-iron columns with fluted heads, eighteen to a row and spaced 35ft apart, support the longitudinal cast-iron pierced spandrel arches, which in turn support roof trusses of 32ft span. The ironwork was supplied by J. & R. Mallet, the Dublin iron founders. The general contractor was William Dargan, with Cockburn & Williams being responsible for the main building. A two-storey riverside wing was added in 1911. Between 2002 and 2004, a major refurbishment resulted in the roof trusses of the original shed being replaced in steel. (HEW 3091)

D13. Inchicore Railway Works (O 112335). The original buildings at the railway works at Inchicore in Dublin were built between 1844 and 1846 for the Great Southern & Western Railway (GS&WR). The works were in operation from the early part of 1846. The Inchicore works represent the sole survivor of a number of independent railway works in Ireland, and one of the earliest such works surviving in the world. The works were built on an 87-acre estate beside the Dublin–Cork main line, about 2 miles to the west of the terminus of the GS&WR at Kingsbridge (now Heuston) Station (**D12**). The main building was designed by the architect Sancton Wood, who was also responsible for the headquarters of the company at Kingsbridge. The original works included a running shed, two erecting shops, boiler, carriage, paint and wagon shops, smithy and foundry, as well as administration and design offices. The roofs of a number of the buildings are supported by iron roof trusses carried on cast-iron columns. These date from the 1840s and were supplied by the Dublin foundry of J. & R. Mallet and are similar to those to be found at Kingsbridge. The original buildings were built in a very durable blue limestone, the facade towards the railway main line having very distinctive castellated features. (HEW 3203)

Inchicore Railway Works, Dublin

'Loop Line' crossing Westland Row *c.* 1880

D14. Liffey Rail Viaduct (O 167340 to O 167350) forms part of the mile-long 'Loop Line' built in 1884 for the City of Dublin Junction Railway. This connects the Dublin, Wicklow & Wexford Railway terminus at Westland Row with the Great Northern Railway (Ireland) (GNR (I)) terminus at Amiens Street (now Connolly) Station, where a new set of platforms was built alongside those of the GNR (I).

The viaduct crossing the River Liffey consists of a three-span bridge of twin wrought-iron latticed girders supported on pairs of cylindrical section cast-iron caissons sunk to rock and infilled with concrete. The spans from the George's Quay side of the river are 116ft 5in, 131ft 2in, and 139ft 5in respectively. The simply supported main girders, spaced 30ft apart, are 13ft deep with 3ft-wide flanges, the webs being doubly redundant lattices composed of members of varying section. The cross girders originally carried a steel trough deck, but this was replaced in 1958–60 by a system of steel stringers welded to flat deck plating. The twin river piers are cross-braced above high-water level by semicircular arched members with hollow spandrels and are capped by ornamental sections, which carry the main girders. The designer was John Chaloner Smith, and the general contractor M. Meade & Sons of Dublin, the structural ironwork being supplied and erected by William Arrol of Glasgow. The northern approach viaduct consists of similar type, but smaller, girders carried on white limestone piers (supposedly in an attempt to harmonise with Gandon's adjacent Custom House). There are also a number of substantial iron bridges, supplied and erected by A. Handyside & Co. of Derby spanning the main streets crossed by the line, which was opened on 1 May 1891. (HEW 3036)

D15. Butt Bridge (O 162 345). Located immediately upstream of the Liffey Viaduct, the present reinforced concrete bridge, which replaced an earlier iron swing bridge, was named after the Irish parliamentarian Isaac Butt. The bridge was designed by Francis Bond under the direction of Joseph Mallagh (chief engineer to the Dublin Port & Docks Board), Pierce Purcell (professor of civil engineering at University College Dublin) acting as adviser to the Board. Bond, who later succeeded Mallagh as chief engineer in Dublin Port, served as resident engineer on the project, and Thomas Simington as site agent for the contractor Gray's Ferro-Concrete (Ireland) Ltd. Opened in 1932, the bridge is 66ft wide and carries a 40ft-roadway and two footpaths of 11ft 9in each. The central span of 112ft is formed by two cantilevered sections, and the two approach spans, each of 39ft 6in, act as counterweights. Over the approach spans the roadway is splayed out to a width of 66ft at the abutments to provide easier access from the quays, the footpaths being carried on ribs cantilevered out from the main structure. The arch profiles are all portions of circular arcs: the central span has a radius of 125ft with a rise of 13ft 3in, while the approach spans have a rise of 8ft 6in and a radius of 27ft 2in on the axis of the bridge, which is slightly skew to the abutments. The parapets are of Wicklow granite. (HEW 3017)

Butt Bridge (left) alongside Liffey Rail Viaduct (right), Dublin

D16. Lucan Bridge (O 035355), the largest single-span masonry arch bridge in Ireland, is to be found where the road from Clondalkin to Clonsilla crosses over the River Liffey at Lucan in County Dublin. Lucan Bridge consists of a single arch of ashlar masonry of 110ft span and 22ft rise, the arch profile being segmental, with a rise-to-span ratio of 0.2 (compared with 0.29 at Lismore **(M6)** and similar to that of Sarah Bridge downstream at Islandbridge). The architect and builder George Knowles was probably inspired by the work of Alexander Stevens at Islandbridge. Assisted by James Savage, Knowles created an aesthetically pleasing bridge with raised voussoirs in a deep arch ring and iron balustraded parapets. The parapets were supplied by the Royal Phoenix Ironworks in Parkgate Street in Dublin; their bases are date-stamped '1814'. There is a long approach to the bridge on the south side of the river, the total length of the crossing being around 500ft. Although there is a weight restriction, the bridge continues to accommodate heavy vehicles from time to time, a good indication of the inherent strength of well-constructed masonry arches. (HEW 3045)

Lucan Bridge

D17. O'Connell Bridge (O 160343). In 1782 the Wide Streets Commissioners obtained a parliamentary grant to erect a bridge to connect Sackville (now O'Connell) Street on the north side of the River Liffey with Westmoreland and D'Olier Streets on the south. The Commissioners employed the architect of the Custom House, James Gandon, to design the bridge. Consisting of three semicircular masonry arches spanning a total of 150ft between the quayside abutments, this bridge had steep approaches and was quite narrow, being only 40ft wide. It was named Carlisle Bridge in honour of the Lord Lieutenant of the time.

O'Connell Bridge, Dublin

By the 1860s, the city had spread eastwards along the quays and some 10,000 vehicles a day were using the bridge. Rapidly developing commerce, a railway system largely centred on Dublin and the general economic improvement in the country all combined to make the bridge totally inadequate. Following a design competition and many arguments, the proposal of Bindon Blood Stoney, engineer to Dublin port, was finally adopted. The plan was to rebuild the eighteenth-century structure in order to provide a level roadway and to increase the width to that of Sackville Street. The method of construction involved using concrete-filled iron caissons down to rock level to extend the piers to support the new work, together with the rebuilding of the existing arches. The contractor appointed, William J. Doherty, had already rebuilt Grattan Bridge (**D18**) further upstream and considerable lengths of quay wall. The resulting bridge has three elliptical masonry arches: the central arch is of 49ft span with a rise of 13ft, and the side arches have a 40ft span and 12ft 3in rise. The length of the bridge between abutments is 144ft and the width between parapets 152ft 8in. The architectural merits of the bridge lie in the fine granite ashlar masonry work, the balustraded parapets and the sculptured decorations. The bridge was opened in May 1880, when it was renamed in honour of the Irish patriot Daniel O'Connell. (HEW 3040)

D18. Grattan Bridge (O 154342), An Act of Parliament in 1811 made the Ballast Board in Dublin responsible for the bridges over the River Liffey upstream as far as and including Victoria (now Rory O'More) Bridge (**D20**), but excluding the Corporation's Liffey Bridge (**D19**). Responsibility passed to the Dublin Port and Docks Board when it was established in 1867. Its chief engineer, Bindon Blood Stoney, was called upon to undertake the rebuilding and widening of the Essex and Carlisle (now O'Connell) river bridges.

Essex Bridge was named after the Earl of Essex, Arthur Capel, the Lord Lieutenant of Ireland between 1672 and 1677. This seven-span masonry bridge was completed in 1678 for Sir Humphrey Jervis and was replaced between 1753 and 1755 by a five-arch masonry

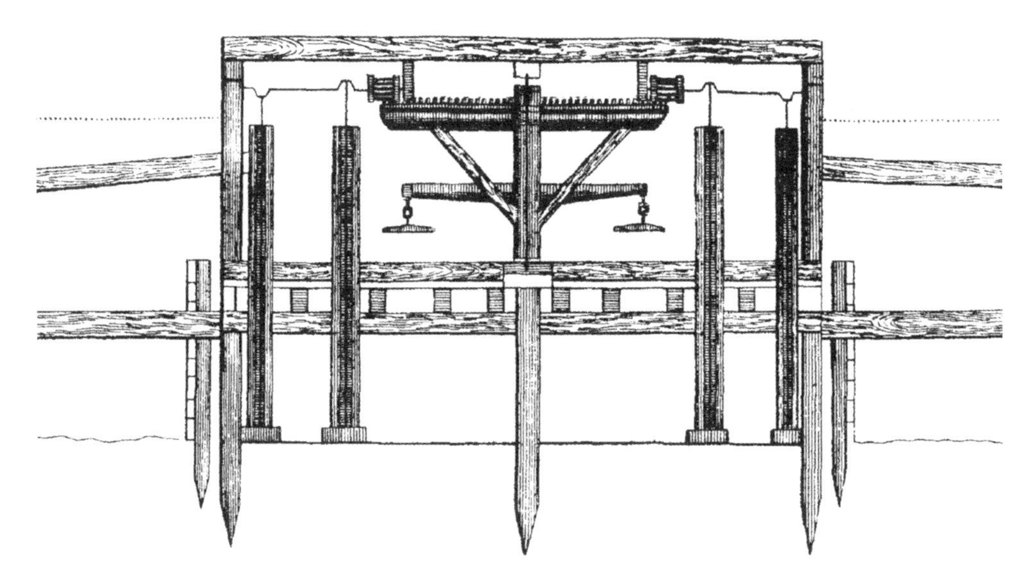

Semple's apparatus for dewatering the foundations of Essex (Grattan) Bridge

bridge designed by George Semple. He based his design on Labelye's bridge at Westminster in London. In 1865, traffic conditions required that this narrow bridge with its steep approaches be rebuilt or replaced. Stoney at first suggested an iron bridge, but also put forward plans for rebuilding, which involved substituting a series of flatter segmental or elliptical arches for the lofty crowns of the existing semicircular arches, and carrying the footpaths beyond the face of the arches on cantilevered iron girders. This latter plan was adopted and the present bridge was completed in 1875 by the contractor W.J. Doherty, with the ironwork supplied by Courtney, Stephens & Bailey of Dublin. The central segmental arch spans 45ft and is flanked on each side by segmental arches of 39ft 6in span and semicircular arches of 14ft 2in span. The rise-to-span ratio of the segmental arches is around 0.22. The footpaths are supported on wrought-iron frames cantilevered out from the faces of the bridge and bolted through to the lowest courses of the new masonry. The parapets consist of wrought-iron latticed girders with cast-iron ornamentation. The lamp supports are in the form of sea horses. Following the rebuilding and widening, the bridge was renamed Grattan Bridge after the parliamentarian Henry Grattan. (HEW 3041)

D19. Liffey Bridge (O 158343). The earliest known iron bridge of any significance in Ireland, the Liffey Bridge was erected in 1816 to connect Merchant's Arch on the south quays with Liffey Street Lower leading from the north quays. The Liffey Bridge is a single span cast-iron arch with an elliptical profile. The bridge consists of three parallel arched ribs spanning 137ft 9in between angled masonry abutments and having a rise of 11ft 9in. (The span increases to about 141ft at deck level). Each arch rib consists of six lengths of cruciform section. These are connected together at each rib joint to form two tiers of rectangular openings with chamfered surround, the depth of the opening decreasing towards the crown. The ribs are stiffened by the deck and by diagonal and normal bracing to form a truss in the plane of the intrados. The transverse members are of hollow circular section with a bolt passing through, and act as spacers to provide lateral stability. Cast corbels on the outside ribs carry a flat plate that supports the parapet railings. The bridge is supported by fish-bellied beams that span perpendicularly to the arch ribs. Three lamps are carried on ornamental supports mounted on top of the railings, their supports providing lateral stability for the railings. The bridge is thought to have been designed by John Windsor, a foreman at the Abraham Darby III foundry at Coalbrookdale in Shropshire. In 2001–02 the bridge was completely refurbished, the ironwork being sent for restoration to Harland and Wolff in Belfast. (HEW 3032)

Liffey (Ha'penny) Bridge, Dublin

Rory O'More Bridge, Dublin

D20. Rory O'More Bridge (O 143343) spanning the River Liffey between Ellis Street on the north side and Watling Street on the south is the third bridge to be erected on or near the site. The first bridge, erected around 1670, was of timber and replaced a ferry. This bridge was replaced in 1704 with a multi-span masonry bridge, which in turn was replaced by a single-span iron bridge opened in 1861. This is now known as Rory O'More Bridge (after one of the leaders of the 1641 rebellion).

The bridge has seven segmental cast-iron ribs of I-section at 5ft 6in centres, spanning 95ft on a slight skew between granite ashlar abutments at the quay walls, the rise of the arches being 9ft 6in. The ribs are 24in deep with 9in-flanges and there is wrought-iron diagonal bracing between the ribs. The carriageway is 23ft wide with 5ft-wide footpaths each side. The decking is carried on buckled plates spanning between I-beams placed on top of the spandrels, which are composed of a number of vertical members connected at their tops by semicircular arched elements. The arches spring from masonry abutments. The ironwork was supplied by Robert Daglish junior of the St Helen's Foundry in Lancashire, the general contractor being John Killeen of Malahide. According to the inscription on the arch, the ironwork was cast in 1858, although the bridge was not opened until some three years later. (HEW 3039)

D21. Sean Heuston Bridge (O 138343), spanning the River Liffey near Heuston Station, is an iron road bridge erected in 1828 to commemorate the visit of George IV to Ireland seven years previously. The bridge, formerly called King's Bridge, but now named after an Irish patriot, Sean Heuston, consists of a single cast-iron span of 98ft with a rise of 16ft 10in. There are seven ribs, the outer ones, together with the spandrels and parapets, being highly ornamented. The bridge was designed by the architect George Papworth and the castings produced by Richard Robinson at the Royal Phoenix Iron Works in nearly Parkgate Street. The bridge has an overall width of 31ft 2in. The bridge was strengthened in 2003 by the replacement of the inner cast-iron ribs with steel beams to carry a LUAS light-rail line across the river. (HEW 3033)

Sean Heuston Bridge, Dublin

D22. Donnybrook Bus Garage (O 178313). The main building at the bus garage at Donnybrook in south Dublin was completed in 1951 and is covered by the most extensive area of concrete shell roofing in Ireland. Columns support a roof of ten thin slab vaults. The width of the structure is 110ft and the length about 400ft, divided into two five-bay sections. Each vault is 40ft wide and spans 104ft between the main columns. The height to the underside of the longitudinal beams is 20ft. The vaulted roof springs from a point 26ft 2in above floor level and the highest point of the roof is 33ft 6in. Roof lights, 7ft 6in wide, extend the full length of the vaults. The external walls are secured to the columns by flexible ties to give the roof complete freedom to expand longitudinally. The vaulted roof slabs are only 2½in thick, splaying out to 5in for a short distance at the springing, at the top light trimmer beams, and at the junction with the gable walls. There are five layers of reinforcing steel of varying diameter with ½in cover top and bottom. The architects were Michael Scott and James Brennan and the structural design was by Ove Arup & Partners. The contractors were H.C. McNally & Co., in collaboration with Messrs Larsen and Neilsen A.S. of Copenhagen. (HEW 3224)

Donnybrook Bus Garage, Dublin

D23. Balrothery Weir (O 113277). As early as the thirteenth century, efforts were being made to ensure a better and more reliable water supply to Dublin city. At that time, the main source of supply to those living within the city walls was from the River Poddle (now entirely culverted). The somewhat limited catchment area of the Poddle, a tributary of the River Liffey, lies to the west of Tallaght. In the thirteenth century, the monks of the Priory of St Thomas constructed a weir across the River Dodder at Balrothery near Firhouse at which point they abstracted a water supply for the priory. In order to increase the flow in the River Poddle, and so improve the water supply to Dublin, the weir at Balrothery was raised in 1244 and water diverted through an open artificial channel some 1¾ miles long to the Tymon River upstream of Kimmage Manor at Whitehall Road (from here the Tymon River formed a natural channel to the River Poddle). The flow in the canal was controlled by four sluice gates. The weir is about 230ft wide and 16ft high and is built of rough limestone blocks. It was reconstructed by Andrew Coffey in the early nineteenth century, the present sluice gates and bypass channels dating from that time. Major repairs to the weir were also required following damage caused by Hurricane Charlie in 1986. (HEW 3229)

Balrothery Weir near Dublin

Pump House at Stillorgan Reservoirs, Dublin

D24. Vartry Aqueduct and Stillorgan Reservoirs (O 215014 to O 200268). The original Vartry water supply scheme, built between 1862 and 1867, included one large service reservoir at Stillorgan at the end of an aqueduct from Roundwood in County Wicklow. The aqueduct is 33 miles long. It begins as a 6ft high by 5ft wide tunnel, which was driven in solid rock for 2½ miles, and continues as a twin pipeline. Two different early tunnelling machines were used for driving part of the tunnel but the rock was found to be too hard and the tunnel was largely dug by hand and is unlined.

The aqueduct is carried over the Dargle and Cookstown rivers on ornamental iron-latticed truss bridges of 65ft and 45ft span respectively. A valve house and screen chamber were built at the northeast corner of the service reservoir, but these are no longer used. The extension of the Vartry supply during the period 1908 to 1925 included the provision of two new reservoirs at Stillorgan and a new valve house. The Stillorgan reservoirs were created partly by excavation, partly by earthen embankments, the highest embankment being about 26ft above the general ground level. The Lady Gray reservoir at Stillorgan, completed in 1923 has a capacity of 90 million gallons.

The original valve house, which contains the screen chamber, is an octagonal-shaped building with sides of 18ft. The building rises about 15ft above ground and extends to 50ft below ground. There are some fine decorated iron trusses supporting the roof. The design of the original scheme was the responsibility of the Dublin City Engineer, Parke Neville. The pipework, valves and other ironwork were supplied and erected by Edington & Son of Glasgow. (HEW 3022 and 3023)

D25. UCD Water Tower (O 178312). A very striking reinforced concrete water tower was built on the Belfield campus of University College Dublin (UCD) in 1969–70. The tower was provided to ensure an adequate supply at a pressure higher than what could, at the time, be provided from the public mains. The tower is no longer used for this purpose.

Facing page: Water tower on campus of University College Dublin

The top of the tower is 184ft above ground level. The shaft is a regular pentagon with an outer side length of 7ft 6in. The wall thickness is 20in for just over the first third, when it reduces to 14in. The 150,000-gallon asymmetrical pentoidal tank with a maximum width of 41ft 6in. The original design comprised a dodecahedron with twelve pentagonal faces, but the lower part had to be modified to allow for the pipework, and for merging into the base. The tank walls are also 14in thick. In line with one side of the main shaft, a 5ft by 3ft access shaft passes through the tank to the flat roof. The whole shape has been emphasised by 2in deep fluting. The base comprises a 5ft-thick 44ft-square reinforced concrete raft. The architect was Andrew Wejchert, the structural design by Thomas Garland & Partners, and the contractor Paul Construction Ltd of Dublin. (HEW 3240)

MUNSTER

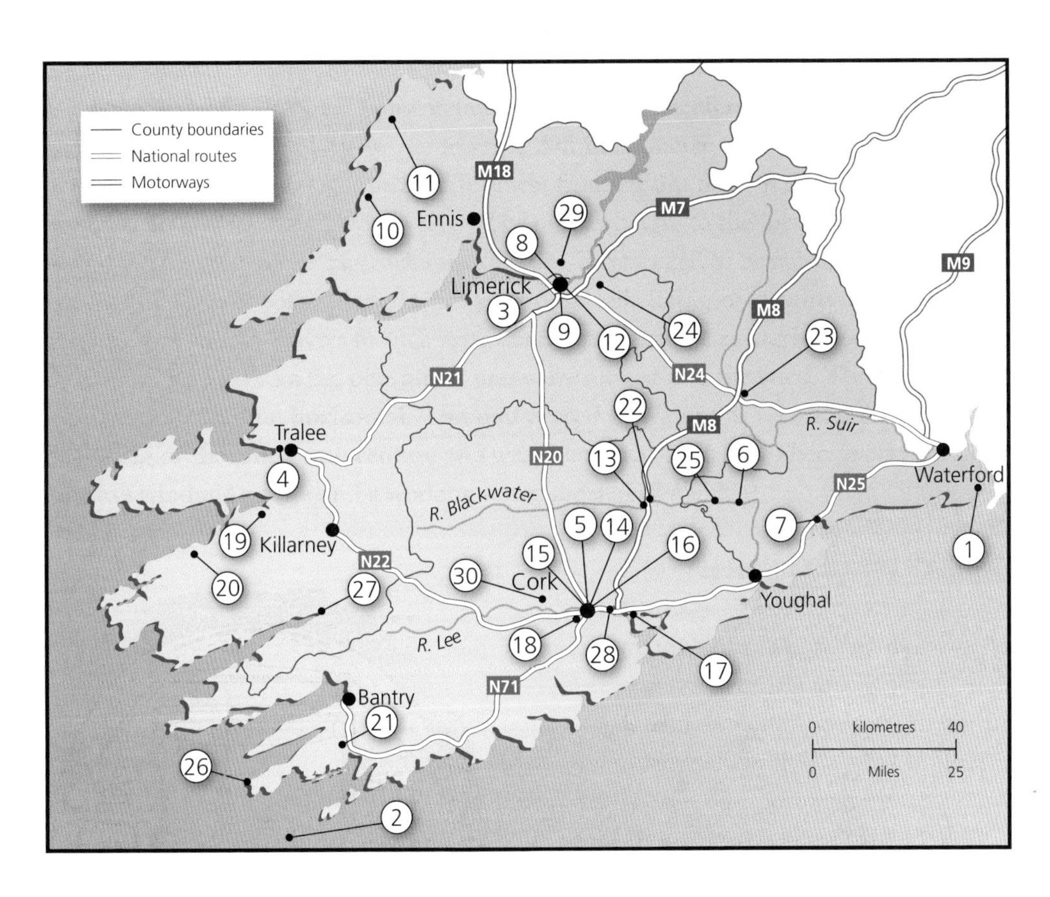

M1 Dunmore East Harbour and Lighthouse
M2 Fastnet Rock Lighthouse
M3 Limerick Docks
M4 Tralee Ship Canal
M5 Lee Road Waterworks, Cork
M6 Lismore Bridges
M7 Causeway Bridge, Dungarvan
M8 Thomond Bridge
M9 Sarsfield Bridge
M10 Bealaclugga Bridge
M11 Spectacle Bridge
M12 Athlunkard Bridge
M13 Fermoy Bridge
M14 St Patrick's Bridge, Cork
M15 Daly's Bridge, Cork

M16 Rathmore Rail Tunnel, Cork
M17 Slatty and Belvelly Rail Viaducts
M18 Chetwynd Rail Viaduct
M19 Laune Rail Viaduct
M20 Gleensk Rail Viaduct
M21 Ballydehob Rail Viaduct
M22 Carrickabrack Rail Viaduct
M23 Suir Rail Viaduct, Cahir
M24 Barrington Bridge
M25 Ballyduff Upper Bridge
M26 Mizen Head Bridge
M27 Our Lady's Bridge, Kenmare
M28 Jack Lynch Road Tunnel, Cork
M29 Shannon Power Scheme, Ardnacrusha
M30 Inniscarra Dam

Dunmore East Harbour

M1. Dunmore East Harbour and Lighthouse (S 690000) lies on the southeast coast of County Waterford at the entrance to Waterford Harbour. The original pier and lighthouse were built under the direction of the civil engineer Alexander Nimmo, between 1823 and 1825. The harbour was built for the British Post Office to accommodate the mail packet ships operating out of Milford Haven in South Wales (later steam ships were transferred to Waterford).

The pier is 780ft long and 77ft wide at the base. A substantial parapet wall rising to a height of 14ft above the surface of the pier provides shelter from easterly gales, the seaward side of which is protected by battered masonry slopes or pavement. The pier terminates in a bull-nosed platform on which stands the lighthouse, a fluted Doric column of Old Red Sandstone conglomerate 50ft high.

The pier was partly rebuilt in 1832 by the Board of Works under the direction of Barry Duncan Gibbons and a further 16ft-wide breakwater, extending 295ft to the northwest from the end of the pier, was added in the 1960s. Also at this time, an extra storm wall, extending for a length of 666ft, was constructed over the original pier parapet. It is 10ft 4in wide at the crest, about 16ft wide at the base, and incorporates a curved wave deflector. Between 1963 and 1972, Dunmore East Harbour was developed further as one of five national fishery harbours. (HEW 3072)

M2. Fastnet Rock Lighthouse (V 882163) stands on a jagged pinnacle of rock, $4\frac{1}{2}$ miles off the Cork coast southwest of Cape Clear, and is the most iconic of all Irish lighthouses.

Fastnet Rock Lighthouse JOHN EAGLE PHOTOGRAPHY

The first lighthouse tower on the Fastnet Rock, designed by George Halpin senior, the engineer to the Ballast Board in Dublin, was constructed between 1848 and 1853. The cast-iron tower, supplied and erected by J. & R. Mallet of Dublin, was 63ft 9in high, tapering from a diameter of 19ft at the base to 13ft 6in at the top. It consisted of flanged plates bolted together, the base being secured by long bolts into the rock surface. The base was filled solid with rubble masonry to a depth of 3ft and lining walls of masonry 3ft 6in thick were built up to first floor level. The lantern added another 27ft 8in to the overall tower height of 91ft, the light itself being 148ft above sea level. The tower was modified in 1867 to make it more resistant to heavy sea conditions, but doubts were raised regarding the stability of such towers, and it was decided by the Commissioners of Irish Lights in 1891 to build a completely new lighthouse on the rock. The old tower was taken down as far as the room immediately above the solid base section, the room subsequently being converted into an oil storage tank for the new lighthouse.

Construction of the present tower, designed by William Douglass (Engineer-in-Chief to the Commissioners), was commenced in 1896 and continued, as weather permitted, until June 1904, when the powerful light was exhibited for the first time. The tower is built from Cornish granite, supplied by John Freeman & Co. of Penryn, each of the granite blocks being dovetailed into adjacent blocks in each ring of masonry. The tower contains 58,000cu.ft of masonry, the 2,074 blocks weighing in total 4,300 tons. The base diameter of the tower is 52ft tapering to 40ft at a height of 20ft above the base. Partial rings in this section form the facing to the natural rock face. The total height of the masonry tower is 146ft 4in, the external profile being a segment of an ellipse up to a height of 116ft, and vertical thereafter. Including the lantern room, the overall height of the lighthouse is 179ft 6in, the light being 157ft 11in above high water level. The operation of the light became automatic in 1989. (HEW 3049)

M3. Limerick Docks. The early development of Limerick as a port included the building of quays on the west bank of the Abbey River and Arthur's Quay and Honan's Quay on the south bank of the River Shannon. At the time of building Wellesley (now Sarsfield) Bridge in the 1820s, the intention was to form a wet dock upstream of the bridge with access provided by a lie-by and locks. Instead, the quays were extended westwards along the south bank of the river, but no storm protection was provided for shipping, so in 1823 it was decided to construct a wet dock at the western extremity of the then line of quays.

The dock (R 567567), as originally built, was 800ft long by an average width of 390ft, with a 70ft-wide tidal entrance lock at the northeast corner. The work was at first carried out by a contractor named Lawrenson, but was completed in 1853 under the direction of John Long of the Board of Works using direct labour, some thirty years after the project was first mooted. The lock gates were supplied by J. & R. Mallet of Dublin, but were

Limerick Wet Dock GARETH COX

replaced in 1888 with steel cellular gates by Kincaid & Co. of Greenock. The original dock enclosed about 7½ acres and provided around 2,300ft of wharfage.

A large graving dock, 428ft long overall by 397ft wide over the blocks, was constructed in 1873 at the eastern end of the dock with an entrance width of 45ft. The limestone masonry work is particularly fine, with some of the sections of dock being cut directly into the limestone bedrock. The original wrought-iron floating caisson gate by George Robinson & Co. of Cork is still extant, although badly corroded. Power for dewatering the graving dock was originally provided by a steam engine. There is an impressive clock tower, which contained the hydraulic equipment associated with the operation of the graving dock.

The river wall was rebuilt in 1888 by James Henry & Sons of Belfast under the direction of William J. Hall. The wet dock was extended westwards in 1937, thereby increasing the area to around 11 acres and the length of wharfage to around 3,000ft. A new, more convenient, entrance was provided in 1956 at the northwest corner of the docks, the original gates being removed and the northeast entrance filled in. (HEW 3011, 3012)

M4. Tralee Ship Canal (Q 803137 to Q 830140) Plans for the construction of a ship canal to give work to the poor and to improve access to Tralee in County Kerry from the sea were first drawn up by the government engineer Sir Richard Griffith in 1828 and an

Tralee Ship Canal at Blennerville BRIAN GOGGIN

Act obtained the following year. The design was originally for an impounded canal supplied from springs, but was later altered to provide a tidal canal with an increased depth and width. A number of local contractors were employed but progress was slow, so the Board of Works took over and the greater part of the design can be attributed to its engineers, Buck and Owen. The canal finally opened to shipping on 15 April 1846.

The main waterway of the canal is about 74ft wide at high tide and about 15ft deep and was intended to accommodate sailing ships of 200–300 tons displacement. The bed and sides are paved throughout to a roughly parabolic section with local limestone slabs. The canal is slightly less than 2 miles long and commences at a 30ft-wide sea lock (refurbished in 2001–02) at Annagh on the north side of the estuary of the River Lee opposite the restored Blennerville windmill. A local road is carried over a swing bridge (renewed 2001). There are lie-bys at this point and beside the sea lock. The canal terminates at a dock or basin 400ft by 150ft by 15ft deep, which was filled in following the abandonment of the navigation in 1945, but both the canal and basin have recently been restored for occasional use by pleasure craft. (HEW 3062)

M5. Lee Road Waterworks, Cork (X 047990). Until the 1980s, the main water supply for the city of Cork was abstracted from the River Lee upstream of a weir at Lee Road, about 2 miles west of the city centre. Following treatment, the water was pumped to a series of reservoirs on the hills to the north of the city and then fed to the mains. In 1768, the Cork Pipewater Company (established in 1761) installed the first waterwheel and pumps at the Lee Road works to the plans of Davis Ducart. Cork Corporation took over responsibility for the works in 1856 and the City Engineer, Sir John Benson, designed a major scheme of improvements, including the installation of a number of reciprocating pumps driven by water turbines operated by the river flow. The original turbines were installed in 1863 on the site of the present turbines. Three new turbines were installed in 1890 and two more in 1895. Four of the five turbines were supplied by an American company, Jas. Leffel of Ohio. Between 1912 and 1920, new turbine pumps were designed by the waterworks engineer, Michael Riordan, the castings being prepared by the Cork firm of Robert Merrick.

The turbines, and their associated reciprocating pumps, are housed in a fine turbine house. They are preserved, along with some interesting emergency standby steam plant, installed by Combe Barber & Co. of Belfast between 1905 and 1907, consisting of three triple expansion engines. The pumps at the Lee Road plant are now operated by electricity. Until 1875, untreated river water was used to supply the city. In that year, a filter tunnel was laid down in the gravel beds located parallel to the riverbank. It was designed as an infiltration gallery and consists of a 42in-diameter brick tunnel some 1,500ft in length. The tunnel terminates in a circular pure-water basin from which the filtered water was, until the early 1990s, pumped to the reservoirs. Rapid gravity filters were added in 1928, when chlorination of the water was carried out for the first time, and further filters in 1935. The open reservoirs, in use since the mid-nineteenth century, were replaced in 1966 with

Lee Road Waterworks, Cork COLIN RYNNE

Lismore Bridge approach viaduct

6.6Mg-capacity circular prestressed concrete tanks, and in 1975 a water tower of 0.4Mg capacity was built at Churchfield beside the open service reservoir constructed in the 1940s and the latter replaced with two circular concentric concrete tanks.

Since 1982, the City Scheme has been augmented from a major multimillion-pound Harbour and City Scheme, which takes water (with the agreement of the Electricity Supply Board) from the River Lee upstream of the Inniscarra Dam, and from the River Bandon at Innishannon. Much of the Lee Road waterworks has been restored and opened to the public as an educational and interpretative centre known as the Lifetime Lab. (HEW 3077)

M6. Lismore Bridges (X 047990). The first bridge at Lismore in County Waterford was commissioned by the 5th Duke of Devonshire of nearby Lismore Castle to carry the Tallowbridge–Cappoquin road over the River Blackwater. It was built between 1774 and 1779 by Darley and Stokes to the design of the architect Thomas Ivory and consisted of a single span of 100ft over the main channel of the river and an approach viaduct of eight smaller spans (land arches) crossing the floodplain on the north side of the river. Following severe flood damage in November 1853, it became necessary to replace most of the viaduct, the main arch and adjacent span on the north side being retained. The new section of the viaduct, consisting of five segmental arches, was constructed by a contractor called Williams, but some structural weaknesses caused it to collapse the day before it was to be reopened to the public.

Causeway Bridge, Dungarvan

The entire viaduct was completely rebuilt in 1858 by C.H. Hunt and E.P. Nagle to the design of Charles Tarrant, county surveyor of Waterford. The present six approach arches range in span from 40ft 9in to 47ft 3in with rises of between 0.30 and 0.35. These six arches are themselves approached at the northeast end by a three-span masonry arch bridge, also built to Ivory's design. The bridges are all constructed from local limestone and the approach viaduct arches in particular are of some architectural merit. When it was built, the arch spanning the main river channel at Lismore was the longest in Ireland. (HEW 3076)

M7. Causeway Bridge, Dungarvan (X 262934). Since 1987, Shandon Bridge has carried the main Waterford–Cork road (N25) over the River Colligan at Dungarvan. The original line of the road crossed the mouth of the river by the Causeway Bridge. The name derives from a short causeway leading from an area known as Abbeyside, on the opposite bank of the river to the town. The building of the Causeway Bridge formed part of the improvements to the town of Dungarvan undertaken at the beginning of the nineteenth century by the 5th and 6th Dukes of Devonshire, whose Irish seat is at Lismore Castle.

The bridge was designed by Jesse Hartley, a Yorkshire-born civil engineer, who married a local Dungarvan girl and went on to achieve fame as the engineer responsible for developing the Port of Liverpool between 1824 and 1860. The architect to the 6th Duke, William Atkinson, designed the bridge and Hartley supervised the construction between 1813 and 1816.

The bridge spans 77ft 6in, the segmental arch having a rise of 14ft to the scroll keystone at the top of the arch. The width between the solid parapets is only 27ft 8in and there is no footpath of any significance. Although suitable limestone was available locally, Atkinson chose to import sandstone from Runcorn in Cheshire. The joints in the spandrel masonry are at right angles to the intrados of the arch and are continuations of the voussoir joints, thus creating a pleasing fan-shaped appearance when viewed from the harbour road. There are similarities with the masonry work found in the Ringsend Bridge at Dublin, which was completed in 1812 and is of similar span. The heavy ornamented abutments feature decorative rectangular panels and the masonry blocks were given a rough or 'rusticated' finish. (HEW 3061)

M8. Thomond Bridge (R 577579). The present Thomond Bridge spanning the River Shannon in Limerick city replaced an earlier masonry arch bridge of fourteen spans reputedly erected in 1210 in the reign of King John. The bed of the river at this point is formed by a rock ledge, which provides a firm foundation for the masonry piers. The designer, James Pain, used the foundations of alternate piers in the old bridge as the foundations for the new piers. Erected between 1838 and 1840 by the Board of Works, Thomond Bridge consists of five segmental arches of 50ft span and 10ft rise, and two flanking segmental arches of 40ft span and similar rise. There is a generous width of 36ft between the solid parapet walls. Pointed cutwaters are used to direct the flow of the river between the piers. Good quality limestone is to be found locally and this material was used by Pain in Thomond Bridge and in his bridge erected some ten years previously further upstream at Athlunkard (**M12**). (HEW 3059)

Thomond Bridge, Limerick

M9. Sarsfield Bridge (R 574573). Alexander Nimmo, a gifted pupil of the renowned civil engineer Thomas Telford, designed and supervised the construction of many roads, bridges, and piers, mostly in the west of Ireland between 1824 and 1832. One of the best examples of his work is Wellesley (now Sarsfield) Bridge over the River Shannon in Limerick city. Nimmo's design for Wellesley Bridge was greatly influenced by the Pont Neuilly, designed by Jean Perronet, head of the Corps des Ponts et Chaussées, and built across the River Seine in Paris in the 1780s. The foundation stone of the Limerick bridge was laid on 25 October 1824, the contractor being Clements & Son. Towards the end of the construction period, the works were taken over by the Board of Works and in 1831 John Grantham replaced Nimmo as engineer.

The main bridge between the river island and the west bank (R 574573) comprises five arches of limestone ashlar masonry, each of 70ft span and 8ft 6in rise, a rise-to-span ratio of only 0.12. The height above the waterline at which each of the main arches springs reduces from the face of the arch to the bridge centre line. This was done to aid the hydraulic flow at times when high-tide levels coincided with the river being in flood, a condition now controlled to a large extent by the requirements of the hydroelectric power station upstream at Ardnacrusha. The tops of the cutwaters are carved to resemble seashells.

The bridge cut off access from the sea to the existing harbour and quays. Provision was made for shipping to reach the upper city quays and the Shannon Navigation through a lie-by and entrance channel, spanned by a swivel bridge. The original twin-span swivel bridge supplied by Forrester & Co. of Liverpool was replaced in 1923 by a single-leaf swivel bridge from the Cleveland Bridge and Engineering Co. of Darlington. The opening mechanism of this bridge was electrified in 1926 and the bridge fixed permanently in a closed position in 1963. (HEW 3058)

M10. Bealaclugga Bridge (R 038772). During the 1820s, the government of the day, then based in London, carried out extensive public works in Ireland, including the construction of roads and harbours, much of the activity being along the western seaboard. In County Clare, the coast road between Kilkee and Milltown Malbay crosses the Annagh River near Spanish Point. At this location, the government engineer John Killaly erected a masonry arch bridge in 1824, in what might be called the 'castellated Gothic' style, at

Bealaclugga Bridge

the same time exercising a considerable flair for ornamentation. Reminiscent of Alexander Nimmo's bridge over the River Liffey at Pollaphuca in County Wicklow (**L6**), Bealaclugga Bridge, on the N67, has a span of 34ft, the rise to the soffit of the pointed arch being 21ft 9in. The substantial abutments are over 17ft thick and 35ft wide and are heavily ornamented with a mixture of shields, rings and slits. (HEW 3181)

M11. Spectacle Bridge (Q 996332). In 1875, the county surveyor of Clare, John Hill, was faced with the problem of carrying the main road from Ennistymon to Lisdoonvarna over the River Aille, where it flows in a narrow gorge, with near vertical sides, in places up to 80ft high. At the site chosen, the road level is some 46ft above river level. Hill provided a novel design aimed at avoiding excessively deep spandrel walls, thereby reducing significantly the dead weight of the filling. He built an ordinary semicircular masonry arch of 18ft 2in span at low level. Resting on this he constructed masonry brickwork, with a large circular opening of 18ft 2in diameter, which carries the roadway. The visual effect from river level gave rise to the name Spectacle Bridge. The overall length of the bridge is 72ft; the abutments extend 20ft 4in either side of the arch with a width of 24ft 9in and a further 6ft 3in either side with a width of 26ft 9in. Although on a national route, it has to date been found possible to retain this unusual bridge, which is a listed structure. (HEW 3144)

Spectacle Bridge (from an early postcard)

Athlunkard Bridge

M12. Athlunkard Bridge (R 589590). The main road north from Limerick city crosses the River Shannon at Athlunkard. Here, in June 1826, a bridge was built in the massive style to withstand the full force of the river in flood, although river levels are now controlled at Parteen Weir, at the start of the intake canal to the hydroelectric works at Ardnacrusha. A plaque on the bridge records the designers as the firm of James and Geo. Richd. Pain (Architects). The Pains were pupils of the famous London architect, John Nash.

The five segmental arches are each of 67ft span, with a rise of 13ft 6in. The abutments are 30ft thick and the piers each 10ft 6in thick. The width between the solid stone parapets is 26ft, but this widens to just under 50ft at a point some 23ft from the face of the abutments. The bridge is built at right angles to the river and is approached by embankments, 111ft long on the east side and 235ft on the west. On the west or Limerick side there is an attractive tollhouse, which is now a private residence. Athlunkard Bridge was opened in December 1830, having taken four and a half years to complete. (HEW 3131)

M13. Fermoy Bridge (W 812985) in County Cork is a fine example of a mid-nineteenth-century masonry structure built to last. The earliest bridge across the River Blackwater at Fermoy was built around 1625 and had many small arches, which tended to impede the flow of the river. The present bridge, in ashlar limestone with rough-dressed facing, represents the partial rebuilding of an earlier structure. Up until the opening of the M8 motorway, it carried the main Dublin–Cork road over the river on the northern approach to the town. The bridge was completed in 1865 by Joshua Hargreave to the design of A. Oliver Lyons.

Fermoy Bridge

There are seven segmental arches, which vary in span from 38ft 6in to 48ft 4in with rise-to-span ratios of around 0.22. The width between the solid parapets is a generous 41ft, allowing adequate room for modern traffic volumes. Retaining walls contain a 190ft approach embankment at the northern end and there are rounded cutwaters both upstream and downstream.

Fermoy has a long history of flooding from the River Blackwater and a flood alleviation scheme is currently under construction. The scheme consists of a combination of embankments, permanent walls, temporary walls and flood control gates. (HEW 3070)

M14. St Patrick's Bridge, Cork (W 676721). The tidal channel of the River Lee in Cork city is spanned between St Patrick's Quay and Merchants Quay by a fine masonry arch bridge of ashlar limestone. An earlier bridge on the site, erected in 1789, was swept away by the great flood of 1853. The present bridge was erected between 1859 and 1861 and reconstructed in its original form in 1981. It has three arches, having an elliptical or three-centred profile, the arch rings being chamfered to assist water flow. The central span is 62ft 4in, and the two flanking spans each 56ft 9in, with an arch rise-to-span ratio of around 0.24. The bridge is nearly 60ft wide between the parapets, which consist of pierced balustrades. The foundations were taken down to 14ft below low-water ordinary spring tides and the piers were formed from cast-iron caissons filled with concrete, the interior being reinforced with transverse bars. The contractor for the 1859 bridge was Joshua Hargreave, the grandson of the man who erected the 1789 bridge. St Patrick's Bridge was

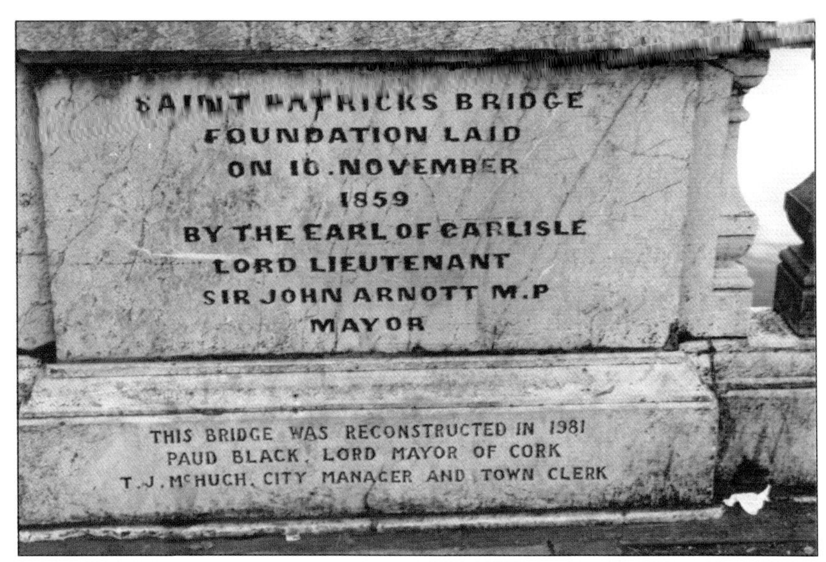

St Patrick's Bridge, Cork (plaque)

designed by the City Engineer, Sir John Benson, who was both a civil engineer and a distinguished architect, and responsible for a number of public buildings and bridges in the city and county of Cork. (HEW 3149)

M15. Daly's Bridge, Cork (W 657716). Pedestrian access to public facilities on the south bank of the north channel of the River Lee at Sunday's Well in Cork city is provided by a striking suspension bridge of 150ft clear span of the piers. Cork Corporation (now the City Council) received the bridge in 1927 as a gift from James Daly of Dalymount in Cork. The 4ft 8in wide walkway was originally of timber and spanned between iron girders, which in turn are supported by vertical suspenders from the suspension cables. The deck has since been concreted. Mass concrete piers of 11ft-square section support the twin towers, 16ft 6in high and 3ft 4in by 1ft 6in in plan. Each tower is formed of angle iron with double-latticed cross-bracing, the towers themselves being cross-braced at right angles to the bridge centre line. The bridge deck is suspended by vertical hangers from two pairs of 1½in-diameter twisted steel cables taken over the tops of the towers to anchorages on each bank. The latticed hand railing, comprising 5ft by 3ft 10in panels, is stayed by brackets carried from the ends of the deck support girders. Daly's Bridge was designed and erected by the London firm of David Rowell, Engineers, for the Cork City Engineer, Stephen W. Farrington. (HEW 3143)

Daly's Bridge, Cork

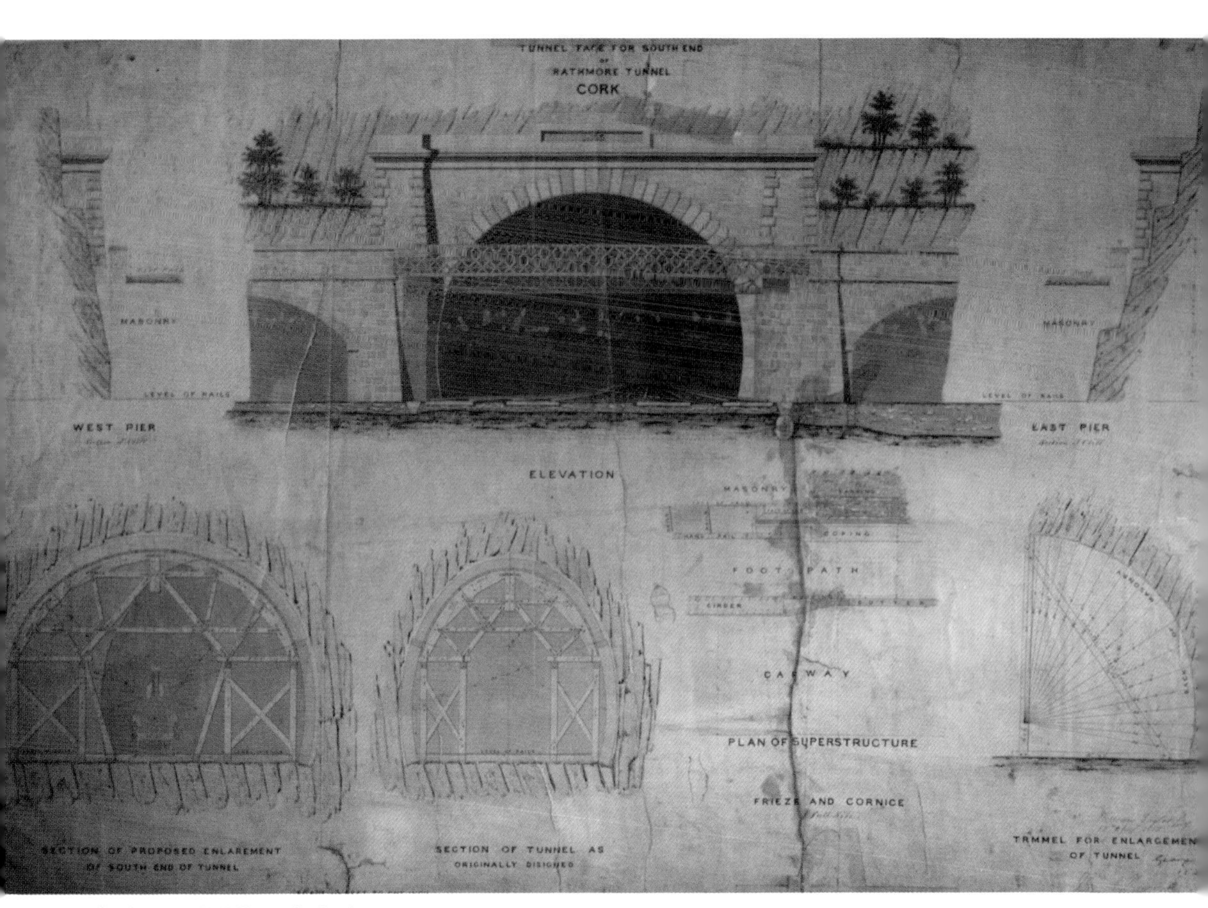

Rathmore Rail Tunnel, Cork

M16. Rathmore Rail Tunnel (Cork) (W 678734 to W 684722). In its heyday, the Great Southern & Western Railway Company (GS&WR) was the largest in Ireland and provided rail facilities between Dublin and the south and southwest of the country. Mallow was reached from Dublin by March 1849 and William Dargan was then entrusted with a single large contract to complete the line to Cork, John Macneill being engineer to the company.

The north Cork countryside is quite hilly and presented much more of a challenge to the railway engineers than had the preceding sections through counties Kildare, Laois and Tipperary. There are, in addition, a number of significant barriers to the progress of the line, one of which was the River Blackwater immediately south of Mallow. The river was spanned here by a ten-arch masonry viaduct, which was blown up during the Civil War and subsequently rebuilt to a different design. The terrain near Kilnap and Monard demanded even higher masonry viaducts and considerable rock cuttings.

The high ground to the north of the River Lee in Cork city posed a major challenge for the contractor and the line had to be terminated in October 1849 at a temporary station (Victoria) in the district of the city known as Blackpool. Macneill decided on a tunnel, to be constructed on a downgrade of 1 in 60, and 4,065ft long, from Blackpool to a terminus at river level in Cork at Penrose Quay. Dargan commenced work in 1853 and the tunnel was opened to traffic on 3 December 1855. The present terminus was completed in 1893 near the entrance to the tunnel at Lower Glanmire Road (now called Kent Station). The rail tunnel, the longest in Ireland still in use for rail traffic, was driven in Old Red Sandstone and brick-lined throughout, except for a short section in the centre. The tunnel cross section is about 28ft wide by 24ft high, except where it widens to nearly double this to allow the tracks to diverge into the terminal station. (HEW 3114)

M17. Slatty and Belvelly Rail Viaducts (W 783719 and W 781710). From 1859, Cobh (previously Queenstown) became an important port of call for transatlantic liners (served by tender to an anchorage in Cork Harbour). Situated on Great Island, Cobh is connected by rail to Cork via a branch leaving the main line at Cobh Junction (now Glounthaune). The branch was opened in 1862, the contractor being Joseph Philip Ronayne of Cork. To reach the island, the line has to cross both the Slatty and Belvelly channels by means of viaducts. The original iron structures were designed by Kennett Bayley in 1887 to replace earlier timber viaducts. The larger of the two, the Slatty Viaduct over the channel between Harper's Island and Fota Island, has six 70ft spans, each comprising a set of three wrought-iron parabolic trusses, the maximum height of the top chord being 11ft. These trusses are carried on 5ft-diameter cast-iron cylindrical piers infilled with concrete. The abutments are pairs of masonry towers founded on timber piles, and are spaced 17ft apart at base level and connected by relieving arches of 13ft 6in span and 3ft rise. Belvelly Viaduct, which was repaired after being seriously damaged in 1922 during the Civil War, has two 60ft spans and one of 70ft, and is of the same general arrangement as its neighbour.

In 2005, Irish Rail undertook major work on both double-track viaducts, with an investment of some €6m. The approach arches and the girders of the Slatty Viaduct were strengthened to accommodate a new 'silent running' deck, in which the rails are embedded in rubber, whilst the spans of the Belvelly Viaduct were replaced in steel. The contractor was Ascon Ltd and the new steelwork supplied by Thompson Engineering of Carlow. (HEW 3111)

Slatty Viaduct IARNRÓD ÉIREANN

M18. Chetwynd Rail Viaduct (W 636677). To the southwest of Cork lies the important market town of Bandon, once a centre for cotton and woollen manufacture. The Cork & Bandon Railway (later to become the Cork, Bandon & South Coast Railway) was completed in 1849 as far as Ballinhassig. This western part of the line was built to the design of Edmund Leahy by four different contractors (Bolton, Henright, Jones & Parrett, and Condor), and included a short tunnel of 510ft at Kilpatrick, the first in Ireland through which trains passed. The final 4 miles of the route into Cork city included the construction of the Chetwynd Viaduct with four spans of 110ft with a headroom of 83ft, and a tunnel of 2,700ft, Gogginshill, just east of Ballinhassig. The viaduct was designed by Charles Nixon, a pupil of Isambard Kingdom Brunel, constructed by Fox Henderson & Company, and opened in 1851. The general contractor was John Philip Ronayne.

The viaduct comprises cast-iron arches, each having four ribs with transverse diagonal bracing and latticed spandrels. The arches span between substantial masonry piers, 20ft by 30ft in section. Cross girders originally carried iron plates that supported the single-track permanent way. The structure was repaired following damage sustained during the Civil War in 1922. The line was closed to traffic in 1961 and the decking removed to discourage access. The Chetwynd Viaduct crosses the main Cork–Bandon road and is a listed structure. (HEW 3073)

Chetwynd Rail Viaduct near Cork

M19. Laune Rail Viaduct (V 780968). In 1881, the Great Southern & Western Railway Company (GS&WR), using powers dating from 1871, decided to construct a branch from the Tralee main line at Farranfore for a distance of 12½ miles to Killorglin. Forming part of the eventual route to Valentia, the branch line ran over fairly level ground, but had to cross the River Laune by a viaduct between steep banks near Killorglin.

Opened for traffic in 1885, the viaduct has three 95ft main spans, each consisting of pairs of twin cross-braced wrought-iron bowstring girders of part N-truss and part latticed members. The tapered piers and abutments are of local limestone, whilst the single arches on each riverbank have red-brick voussoirs. Designed by the engineering section of the GS&WR, the viaduct was built by contractors T.K. Falkiner and S.G. Frazer.

Some corroded structural members were replaced in 1950. The railway closed in 1960 and, following the renewal of the decking and the provision of safety handrails, the viaduct was reopened in late 1993 as a public footpath to form part of a pleasant riverside walk. (HEW 3085)

M20. Gleensk Rail Viaduct (V 580885). The Great Southern & Western Railway Company, having opened up the kingdom of Kerry in the 1850s by their route from Mallow to Killarney and Tralee had, by 1885, completed a branch line from Farranfore Junction to Killorglin. The government of the day, considering the possible use of Valentia as a point of departure for the Atlantic mails, financed, by means of grants and local (Baronial) guarantees, the continuation of the railway at standard gauge down the southern shore of Dingle Bay. At Valentia a ferry operated to Valentia Island, where there was a telegraph and meteorological station.

Opened to Valentia on 12 September 1893, this 16-mile section of the line included some of the steepest gradients and most spectacular engineering works on any Irish railway. Following a 3-mile climb of 1 in 59 to Mountain Stage, the permanent way was set on a narrow shelf cut into the hillside with the county road and sea far below. There were a number of tunnels and heavy earthworks, including large retaining walls, and a curving viaduct at Gleensk. The line was closed to traffic in February 1960.

The viaduct, designed by A.D. Price and built by T.K. Falkiner, crosses over the valley of the Gleensk River between Glenbeigh and Cahirsiveen at a height of 70ft and is on a 650ft radius curve, the sharpest on any standard gauge track in the country. The viaduct has eleven spans of 58ft, which comprise steel girders carried on slender masonry piers.

Gleensk Rail Viaduct

Because of the tight radius, the wheel flanges of the rolling stock of passing trains ran along the sides of the rail. This generated considerable heat and water troughs were provided to cool the flanges. (HEW 3071)

M21. Ballydehob Rail Viaduct (V 990355). West of Skibbereen in County Cork lies a remote area leading down to the Mizen Penisula. To serve part of this area, the West Carbury Tramways & Light Railways Company, with the aid of government grants and local (Baronial) guarantees, built a 3ft-gauge line from the terminus of the Cork, Bandon & South Coast Railway at Skibbereen as far the fishing village of Schull. Construction began in 1884 and the line opened to traffic in July 1886. The only major civil engineering work on the line was the twelve-arch viaduct that carried the single-track line over an inlet of Roaringwater Bay near Ballydehob. The semicircular arches, with 24in-deep arch rings of rough-hewn coursed limestone, span between mass concrete piers with coursed rubble facings. The piers are 4ft 3in by 14ft at the base, tapering to 12ft 6in at the springings.

Ballydehob Rail Viaduct

Carrickabrack Rail Viaduct

The 18in-wide parapet walls were subsequently raised to a height of 3ft 9in for safety reasons, the 9ft 4in-wide viaduct now forming a pedestrian walkway linking the two sides of the inlet. Circular walks have been made possible by providing access across a weir built downstream of the viaduct in order to maintain the water level in the estuary. The designers of the viaduct were S.A. Kirkby and J.W. Dorman, and the contractors were McKeon, Robinson & d'Avigdor of London. The line was closed in 1947. (HEW 3106)

M22. Carrickabrack Rail Viaduct (W 825991). The Fermoy & Lismore Railway was carried over the River Blackwater east of Fermoy in County Cork on a viaduct consisting of a main river span and five side spans on the northern side of the river. This line became known as the 'Duke's Line', as it was almost totally financed by the Duke of Devonshire of Lismore Castle as a means of improving the transport links to his estates and for the use of the local populace. The 15¾ miles of railway, authorised in 1869 and built by John Bagnell in 1872 under the direction of James Otway, was taken over by the Waterford, Dungarvan & Limerick Railway in 1893. This company was in turn absorbed by the Fishguard & Rosslare Railway & Harbours Company, who were responsible between 1903 and 1906 for renewing the viaduct. Only the piers and abutments date from 1872. The central span of 157ft consists of twin 15ft-deep wrought-iron latticed girders with 24in-wide flanges. The five approach spans on the Fermoy side of the river are reversed parabolic steel plate girders with rolled I-section cross girders, the average span being 49ft. (HEW 3115)

Suir Rail Viaduct (Cahir)

M23. Suir Rail Viaduct, Cahir (S 050251). The rail route from Limerick to Rosslare Harbour via Waterford was built in stages between 1848 and 1906. The section from Tipperary to Clonmel was constructed for the Waterford & Limerick Railway (W&LR) between 1851 and 1853 by contractor William Dargan. In order to cross the River Suir at Cahir, the chief engineer to the W&LR, William Richard Le Fanu, specified wrought-iron box girders, spanning between 16ft-thick limestone masonry abutments and two 10ft-thick limestone masonry river piers. The clear spans are 52ft, 150ft and 52ft respectively, the girders being joined together so as to act continuously over the piers. The box girders, made up of plates and angle irons, were supplied by William Fairbairn of Manchester, the patentee for such girder bridges.

The box girders are 11ft 7in deep overall (a span/depth ratio of 13 to 1), 2ft 6in internal width (3ft wide across the flanges), and are spaced 26ft 6in apart. To provide extra resistance to the compressive forces, the top flange is in the form of a box, 1ft 3in deep and 2ft 6in wide, running the length of the main girders. The box girders are divided internally into cells by vertical plates at 2ft intervals.

The permanent way was originally supported on transverse beams spanning between the bottoms of the main box girders. The viaduct was designed for double-line working, but the line was made single in 1931 for reasons of economy. Eighteen of the transverse beams at the Waterford end of the bridge were replaced in mild steel in 1956, following accidental damage. A second accident in October 2003 caused major damage to the permanent way and transverse beams. Reconstruction consisted of fitting new steel floor girders, which are supported from the tops of the original main girders, and the laying of a new concrete deck.

This fine example of a Fairbairn patent-type railway bridge of large span remains in daily use for both passenger and freight traffic. A smaller three-span example near Ballinasloe in County Galway continues to carry traffic on the Dublin–Galway main line over the River Suck. (HEW 3108)

Barrington Bridge, County Limerick (the old and the new)

M24. Barrington Bridge (R 680549). One of the few surviving early iron bridges in Ireland, Barrington Bridge carried the Annacotty–Cappamore road over the Slievenohera River southeast of Limerick city. The bridge was erected in 1818 by J. Doyle of Limerick for Sir Matthew Barrington to gain access from his residence to the parts of his estate situated on the far side of the river. The cast-iron bridge spans 53ft 9in between masonry abutments with a rise of 3ft 1in. The outer ribs are integral with the spandrels and the arches so formed carry a repeated four-leaf clover ornamentation. Immediately above the intrados of the arch there is a pierced design of alternate diamond and oval shapes. Each of the inner ribs unusually consist of an assembly of nine 6ft long, 9in diameter, cast-iron pipes connected longitudinally by gasket-type flanges at the ends of each pipe. The resulting ribs are tied transversely to the spandrels by ties located above the pipes. The width of the bridge is only 13ft 6in and this fact, coupled with a restrictive 16-ton weight limit, led to a new reinforced concrete road bridge being constructed alongside to bypass the earlier structure. (HEW 3060)

M25. Ballyduff Upper Bridge (W 964994). In the 1850s, a timber bridge was erected across the River Blackwater near the village of Ballyduff Upper in County Waterford some 6 miles west of Lismore to connect the Fermoy–Cappoquin and the Fermoy–Lismore main roads. This bridge was replaced in 1887 by the present structure, which consists of twin wrought-iron double-latticed girders spanning from abutments of local dressed limestone either side of the river to a central masonry pier, with spans of 99ft and 103ft respectively. The girders are joined together in order to act continuously over the central pier. The ironwork was supplied and erected by the firm of E.C. & J. Keay of Birmingham, the design being to the requirements of the

Ballyduff Upper Bridge

Mizen Head footbridge (2010 reconstruction) JOHN EAGLE PHOTOGRAPHY

county surveyor, W.E. L'Estrange Duffin. The main girders are 6ft 9in deep, giving a depth to span ratio of approximately 1 to 15. The bridge deck is carried on cross girders, the overall width of the carriageway being only 20ft. The river pier is 6ft wide with a pointed cutwater upstream. The northern approach to the bridge is on an embankment contained within flanking retaining walls, which include a small flood arch. The bridge at Ballyduff Upper represents an important element of Ireland's extant civil engineering heritage. (HEW 3065)

M26. Mizen Head Bridge (V 734234). In order to gain access to the lighthouse station on Mizen Head on the southwest coast of County Cork, a footbridge was constructed in 1909 for the Commissioners of Irish Lights across an inlet to connect the mainland with Cloghan Island. The design chosen consisted of a reinforced concrete through-arch type bridge. The bridge as built spanned 172ft at a height of 150ft above mean sea level. The main ribs had a parabolic curved profile and a trough section and were cast on the hillside before being slung into position by means of an aerial ropeway. The load-bearing trestles and beams were also cast on shore. The bracing was fixed and the ribs filled in, followed by the fitting of the trestles and beams. The floor and hangers were shuttered and cast *in situ*. The rise of the arch ribs was 30ft and the ribs were spaced at 15ft centres at the springing, reducing to 5ft 6in spacing at mid-span. The footway was 4ft 6in wide. When built, Mizen Head Bridge was the longest-span reinforced concrete arch bridge in either Britain or Ireland and the first of its type. The designer was Martyn Noel Ridley of Westminster, London, who used an indented steel bar system of reinforcement, and the contractors were Alfred Thorne & Son, also of Westminster. In 1997, a visitors' centre was opened at the lighthouse station, the bridge providing public access to the centre, which displays early photographs of the bridge and drawings of a number of the alternative designs.

Although repairs were carried out to the bridge in 1973, the severe marine environment continued to affect parts of the structure and, in 2009 it was decided to replace the footbridge with a replica structure, completed in December 2010, thus preserving this landmark feature of the coastline for the enjoyment of future generations (HEW 3026)

M27. Our Lady's Bridge, Kenmare (V 910700). At Kenmare in County Kerry, in 1840, as part of a system of government roads which were being constructed in the southwest of the country, William Bald designed an iron suspension bridge with two half catenaries of pairs of chains. These were suspended from a central tower built on an island in the middle of the estuary of the Roughty River south of the town. This bridge was replaced in 1932–33 by the present reinforced concrete structure designed by Mouchel & Partners and built by contractors A.E. Farr of London at a cost of just over £9,900. The consulting engineer was Pierce Purcell (professor of civil engineering at University College Dublin).

The bridge carries the road between Glengariff and Killarney over the estuary of the Roughty River and comprises two 150ft-span parabolic arches passing through the deck. The arch ribs rise 34ft 4in, are 2ft 6in thick, and 3ft deep at the crown, deepening to 4ft 9in at the springings. The horizontal deck, about 31ft wide, is supported by fourteen transverse beams, 2ft 6in deep by 1ft 3in wide by 29ft 9in long. Ten are suspended from the ribs by hangers at 11ft 3in centres, two rest on short articulated columns near the springings, and two are supported by pairs of columns standing on the rib skew backs at the abutment and central pier respectively.

The present bridge was opened on 25 March 1933. In 1988–89, there was some replacement of damaged concrete and refurbishment of the reinforcement. The structure, originally hinged at both ends, is now fixed at one end and hinged at the other. (HEW3089)

Our Lady's Bridge, Kenmare

Southern approach to Jack Lynch road tunnel, Cork CORK CITY COUNCIL

M28. Jack Lynch Road Tunnel, Cork (W 725722). Named after a former Taoiseach, the Jack Lynch road tunnel allows through-traffic to bypass the centre of Cork city by connecting the southern ring road at Mahon with the Dublin Road (M8) north of the river, providing an interchange with the Waterford/Rosslare Road (N25) at Dunkettle. The overall link is 1.85km in length. The 610m-long reinforced concrete immersed tube tunnel on the bed of the River Lee is made up of five elements, each around 120m long, 24.5m wide and 8.5m high and weighing 27,000 tonnes. These were constructed in a casting basin located partially on the line of the tunnel south of the river at Mahon. The northern approach was formed by a 120m-long floated open 'boat' section – the first of its kind in Ireland. The northbound and southbound lanes are in separate tunnels with a service tunnel in between. A similar tunnel was completed in 2010 under the River Shannon in Limerick city. The design and build main contractor for the Cork tunnel was Tarmac Walls Joint Venture and the consultants were Ewbank Preece O'hEocha, in association with Symonds Travers Morgan. (HEW3255)

M29. Shannon Power Scheme (Ardnacrusha) (R 586618). As Ireland was emerging from the Civil War in the autumn of 1922, a young physicist and electrical engineer, Thomas McLaughlin, joined the firm of Siemens Schukert in Berlin, where he devoted a good deal of his time to developing the concept of harnessing the power of the River Shannon to produce electricity. He was, in time, able to persuade the government of the Irish Free State to accept a plan to construct a single hydroelectric power station at Ardnacrusha near Limerick to achieve the most efficient utilisation of the fall in level

Hydroelectric power station at Ardnacrusha ESB ARCHIVES

between Lough Derg and the Shannon Estuary. Equally important to the success of the Shannon Scheme, as it became known, was to be an electricity grid stretching the length and breadth of the country. Rural electrification was to become the essential framework for the social, economic and industrial development of the country.

The Shannon Scheme, commenced in 1925, involved the construction of a weir and intake near Parteen Villa, head- and tail-race canals spanned by four reinforced concrete bridges, and the power station complex itself. Both the weir and the intake canal are interlinked: the weir across the river regulates the flow of the Shannon and diverts water into the head-race canal, 7½ miles in length, while the flow in the canal is controlled by the intake. The weir was designed to raise the water level in the river at Parteen Villa to that of Lough Derg, thus ensuring that the entire fall between Killaloe and Limerick could be used to drive the turbines in the power station downstream at Ardnacrusha.

Essentially, the power station consists of an intake sluice house, penstocks, generating building, waste channel and navigation locks. The intake sluice house is built on top of a 405ft-long mass-concrete gravity dam across the end of the head-race canal. This regulates the flow of water through the 131ft-long by 19ft 8in-diameter penstock tubes, inclined at a slope of 31°, to the vertical shaft Francis and Kaplan type turbines located in the power house. This is a steel-framed building, the foundations of which lie 98ft below the intake house. Water from the turbines is discharged via 59ft-long draft tubes to the tail-race and is then conveyed back to the river downstream at Parteen-a-Lux. There is a 600ft-long fish pass and navigation locks capable of taking vessels up to 105ft in length.

The Shannon Scheme, one of the largest civil engineering projects of its type in the world at the time it was constructed, was officially opened on 22 July 1929. By 1935,

Construction of dam at Inniscarra, County Cork ESB ARCHIVES

Ardnacrusha was supplying around 80 per cent of the country's electricity requirements. The turbines and generators were subsequently replaced and the capacity of the station increased from its original 82 MW to 110 MW. (HEW 3007)

M30. Inniscarra Dam (W 544722). The first major harnessing of the power of the River Lee for the generation of electricity began in the 1950s with the construction of dams at Inniscarra and Carrigadrohid, upstream from Cork city. The dam at Inniscarra is the only example in Ireland of the buttress type. This design effected a considerable saving in concrete compared to the more normal gravity type dam. It was also claimed that the design reduced the possible uplift on the base of the dam, and permitted a flatter slope to be used for the upstream face, thereby adding a substantial weight of water to the stabilising forces acting on the dam.

The dam is 800ft long and 140ft high and divided into nineteen blocks, mostly 45ft long, with a water-stop between each block. These massive buttressed blocks have 'diamond headed' upstream faces. The buttresses are 18ft wide with a downstream slope of 8:10. The upstream lead of each block is of constant section with a slope of 3:10 and a contact face 6ft 6in wide between blocks.

Upstream at Carrigadrohid, there is a smaller gravity type dam, which together with Inniscarra Dam created two roughly equal storage reservoirs with a combined storage capacity of about 5 per cent of the annual flow of the river. The dams were constructed by the French civil engineering firm Société de Construction des Batignolles. (HEW 3248)

CONNAUGHT

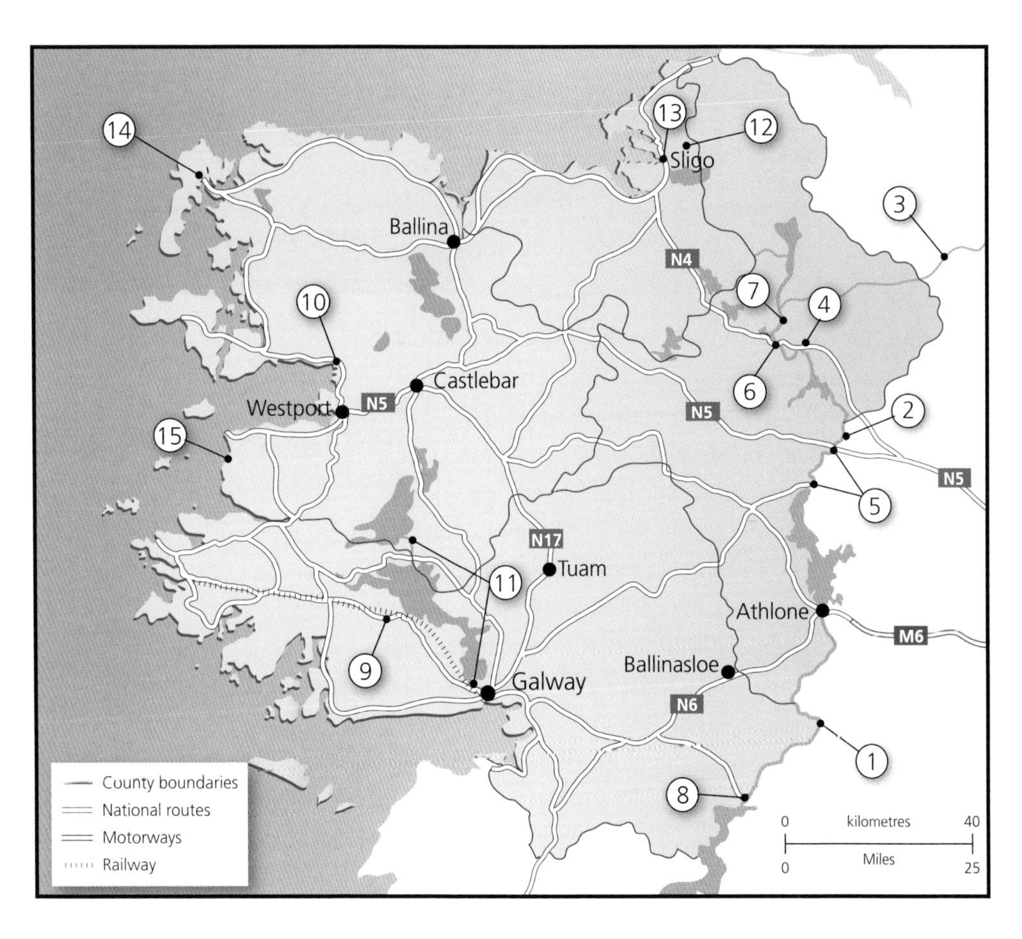

- C1 The Shannon Navigation (Killaloe to Athlone)
- C2 The Shannon Navigation (Athlone to Lough Allen)
- C3 Ballinamore and Ballyconnell Navigation
- C4 Drumsna and Jamestown Bridges
- C5 Lanesborough and Tarmonbarry Bridges
- C6 Carrick-on-Shannon Bridge
- C7 Hartley Bridge
- C8 Portumna Bridge
- C9 Galway–Clifden Railway
- C10 Newport Rail Viaduct
- C11 Eglinton and Cong Canals
- C12 Doonally Reservoir
- C13 Hyde Bridge, Sligo
- C14 Belmullet Ship Canal
- C15 Louisburgh Clapper Bridge

Killaloe Bridge

C1. The Shannon Navigation (Killaloe to Athlone) (R 705730 to N 039416). In the late 1820s, the engineer John Grantham introduced steam vessels to the Shannon Navigation. Other promoters followed suit and steamer passenger and freight services became established, particularly between Athlone in County Westmeath and Killaloe in County Clare, but also as far north as Lanesborough and Carrick-on-Shannon. The Grand Canal (**L8**) from Dublin terminates at Shannon Harbour, about midway between Athlone and Portumna. The remainder of the navigation to Killaloe is through Lough Derg. There are natural obstacles to navigation at four places, namely Athlone, Shannonbridge, Banagher and Meelick, north of Portumna. All the major engineering works required to make the river navigable have been associated with one or other of these locations.

In 1755, the engineer Thomas Omer began work for the Directors General of Inland Navigation and constructed a lateral canal with one lock to bypass the fall in the river at Meelick. The remains of this canal may be seen to the east of the present line of the navigation. Omer also supervised the construction of a canal at Athlone, the line of which may be seen to the west of the present cathedral. It was 1½ miles long and had one lock 120ft long by 19ft wide. From the 1770s onward, the locks and associated engineering works fell into disrepair and, following the Act of Union in 1800, the Directors General of Inland Navigation, acting on the advice of their engineer, John Brownrigg, rebuilt the locks at Athlone and Meelick.

Shannon Commissioners were appointed in 1831 and extensive surveys of the navigation were carried out by their engineer, Thomas Rhodes. A combined drainage and navigation improvement scheme was undertaken in the 1840s, including steam bucket dredging of the main channel. At Meelick, the earlier canal was abandoned and a new large lock 142ft long, capable of accommodating passenger steamers, was constructed in 1845 by the contractor William Mackenzie, and named Victoria Lock. At Athlone, it was decided to deepen the river channel and to construct a substantial weir, 127ft long, and a new lock to replace the earlier bypass canal. The opportunity was also taken to build a new road bridge with an opening navigation span. Large caissons were used to allow construction in the dry of the bridge piers and the lock was built within a large cofferdam. (HEW 3074)

The Jamestown Canal

C2. The Shannon Navigation (Athlone to Lough Allen) (N 039416 to G 962125). The Upper Shannon Navigation extends northwards from Athlone, through Lough Ree, and continues northwestwards towards Carrick-on-Shannon and Lough Key. The navigation is also connected to Lough Allen and Lough Erne. The Royal Canal (**L12**) was opened to Richmond Harbour at Cloondara in County Longford by 1817. It enters the River Shannon near Tarmonbarry to the north of Lanesborough.

Apart from the reconstruction of a number of major road bridges, the main civil engineering work carried out along this stretch of the navigation was confined to the steam-bucket dredging of new channels to bypass the earlier lateral canals constructed under Thomas Omer in the 1760s, the rebuilding of a number of locks and the deepening of the Jamestown Canal south of Carrick. This latter canal had been used by Omer to avoid shallows in the area by cutting across a natural loop in the river. The size of the locks gradually reduces as one moves north, an indication of the perceived commercial potential of the navigation.

The Shannon and Erne systems are now connected by the totally rebuilt Ballinamore and Ballyconnell Navigation (**C3**) (now known as the Shannon Erne Waterway) and represent an important tourism and recreational resource. (**HEW [illegible]**)

C3. Ballinamore and Ballyconnell Navigation (G 957046 to H 353218). Built with the aid of government funding under the Arterial Drainage Act of 1842, the Ballinamore and Ballyconnell Navigation (its original name) commences with a still-water canal of 7½ miles from the River Shannon near Leitrim village to Lough Scur, using a flight of eight locks in a 3-mile stretch. The navigation then comprises a summit-level channel leading from Lough Scur to Lough Marrave, a distance of 3 miles, the remainder of the route to Lough Erne being by way of the canalised Woodford and Yellow rivers. The total length of the navigation is 38½ miles. The original navigation was designed by William Thomas Mulvany (1845) and John McMahon (1846) and was built by direct labour. The navigation was opened in 1860, but was never a commercial success and was closed by 1869.

In 1990, work commenced on an ambitious and expensive project to connect the Shannon and Erne waterways to enhance their tourism and recreational potential. This was a joint venture between the governments of Ireland and the United Kingdom and was also supported by the Ireland Fund and other agencies. The work included the complete reconstruction of all the locks, those at the Leitrim end being 82ft long by 16ft 3in wide with a water depth of 5ft 6in. The locks between Lough Scur and Lough Erne are somewhat wider. All masonry overbridges were either restored or rebuilt, depending on their condition. New steel gates, with timber heel and mitre posts and bottom rails, were fitted to all sixteen locks. The navigation, with its new name of the Shannon–Erne Waterway, was reopened to traffic on 2 April 1994. (HEW 3137)

Reconstructed 14th Lock on the Ballinamore–Ballyconnell Navigation

C4. Drumsna and Jamestown Bridges. (M 993971 and M 981970) Until 2006, the main road from Dublin to Sligo crossed a loop of the River Shannon in County Leitrim at two places, Jamestown and Drumsna.

Drumsna, the older of the two road bridges, erected in the early eighteenth century, has seven segmental masonry arches, increasing in span from 20ft at the west end to 29ft at the centre of the river, again reducing to 23ft at the east end. Another 20ft span arch at the east end was subsequently blocked off. The ring of the most westerly arch was subsequently replaced and the soffit concreted. Several tie bars and pattress plates prevent the spandrel walls from spreading outwards. The pointed cutwaters are taken up to parapet level to form pedestrian refuges. These were deemed to be necessary as the bridge is only 16ft wide between the parapets.

Drumsna Bridge

Jamestown Bridge was completed in 1847 by the Shannon Commissioners under their engineer, Thomas Rhodes. It has five segmental arches of 30ft span and 7ft rise. The large voussoirs in the arch rings are of ashlar limestone, as are the spandrel walls. The cutwaters are rounded, rather than the pointed examples more typical of earlier bridges. The national primary road (N4) now bypasses this part of the River Shannon, thus ensuring that the bridges at both Drumsna and Jamestown will continue their important contribution to the pleasing environment of the Upper Shannon. (HEW 3147 and HEW 3191)

C5. Lanesborough and Tarmonbarry Bridges (N 006695 and N 055770). Work on improving the navigation between Killaloe and Carrick-on-Shannon commenced in 1840. The numerous road crossings of the river were the subject of major engineering works, and in many instances completely new bridges replaced a variety of structures, some timber, some masonry.

At Lanesborough in County Longford, an eight-span masonry bridge was replaced in 1843 with one of six 30ft spans, with a navigation opening of 40ft spanned by a swivel bridge. The swivel bridge was removed in the 1970s and replaced with a fixed reinforced concrete beam and slab arrangement allowing sufficient headroom in the navigation

Tarmonbarry Bridge WATERWAYS IRELAND

Carrick-on-Shannon Bridge WATERWAYS IRELAND

channel. In 1993, a reinforced concrete deck with cantilevered footpaths was built on top of the earlier masonry structure in order to provide the necessary width for the upgraded national primary road between Longford and Roscommon.

At Tarmonbarry the River Shannon is divided into two channels by an island, 85ft wide at the bridge crossing, which is composed of two sections, one of three segmental arches over the east channel and one of four over the west channel, all of 33ft span with a rise of 8ft. The navigation opening on the west side of the river, like that at Lanesborough, was originally spanned by a swivel bridge. This was replaced in the 1970s by the present 40ft-span vertical lifting bridge. The piers are 6ft 6in thick, except for those either side of the navigation opening, which are 20ft. A reinforced concrete deck with cantilevered footpaths was added in 1993 the better to accommodate the main road from Longford to Strokestown. (HEW 3188 and HEW 3190)

C6. Carrick-on-Shannon Bridge (M 938993). During the 1840s, the improvement of the Shannon Navigation required many of the bridges spanning the river to be rebuilt or replaced. Passenger steamers operated between Athlone and Killaloe, but there was less traffic at the northern end of the navigation and the locks were smaller. Opening spans were normally incorporated into bridge crossings, but by the time the improvement works had reached Carrick-on-Shannon, money was running low and it was decided to construct a new fixed bridge at this point with no opening span.

The bridge, designed by Thomas Rhodes, was completed in 1847 and has five spans built of ashlar limestone. These increase from 30ft to 35ft towards the centre of the bridge, with a rise-to-span ratio of 0.23, giving a gradual slope to the parapet. The roadway, however, is level over the central part of the bridge with ramps at each end. There is a smaller arch in the eastern abutment that allows pedestrian movements along the quays. The contractor for the bridge was William Mackenzie. In 2010 a footbridge for pedestrians and cyclists was added to the downstream face of the bridge. (HEW 3146)

Hartley Bridge

C7. Hartley Bridge (G 938022). Financed jointly by Roscommon and Leitrim County councils and the Board of Works, Hartley Bridge was opened in 1915. It carries a minor road over the River Shannon between Carrick-on-Shannon and Cootehall. An early example of the use of reinforced concrete in Ireland, the bridge has 48in-deep parapets that act as beams spanning between supports. The supports consist of pairs of 12in by 12in square columns with cross ties and diagonal bracing struts. The main reinforcement used was of a rail type section, known as a Moss Bar. The spans from centre to centre of the supports are three of 36ft, one of 40ft, and one over the navigation channel of 60ft. The design was prepared for Eugene O'Neill Clarke, County Surveyor of Leitrim and the work was carried out by direct labour. (HEW 3187)

C8. Portumna Bridge (M 871043). A timber trestle bridge was erected across the River Shannon at Portumna in 1795 by the American engineer Lemuel Cox. This was one of seven similar bridges erected in Ireland by Cox. The Portumna Bridge was partially rebuilt in 1818 and replaced in 1834 by another timber structure incorporating an iron swivel bridge over the navigation channel on the Galway side of the river, and causeways forming the approaches to the bridge. This bridge was in turn replaced in 1911 by a steel plate girder bridge.

The Shannon at this point consists of two channels divided by Hayes Island, the one on the north Tipperary side being about 260ft wide, and that on the Galway side being about 240ft wide. Each channel is spanned by three pairs of mild-steel plate girders (either 80ft or 90ft in length) resting on 9ft-diameter concrete-filled cast-iron cylinders. The width

Portumna Bridge

of the approach roadways and bridge is 30ft. The river piers are continued upwards beyond parapet level, tapering to domed tops with decorated finials. The earlier substantial masonry abutments were retained when the replacement bridge was erected. The bridge was designed by C.E. Stanier of London to the specification of J.O. Moynan, the County surveyor of Tipperary (North Riding). The contractors were Heenan and Froude of the Newton Heath Ironworks in Manchester (builders of the Blackpool Tower). The original swing bridge over the navigation channel had unequal arms of 60ft and 30ft length respectively and revolved on a pivotal support on the Galway bank of the river. It was renewed in 2008–09. (HEW 3161)

C9. Galway–Clifden Railway (M 303254 to L 658507). Under the Light Railways (Ireland) Act of 1889, the Midland Great Western Railway Company (MGWR) was provided with a government grant of £264,000 to build a line from their terminus at Galway across the rugged grandeur of Connemara to Clifden. The intention had been to improve communications with a developing fishing industry and the MGWR engineers designed a route to follow the coastline, where the population was estimated to be around 60,000. However, a Royal Commission on Public Works thought otherwise and directed that an inland route should be followed via Oughterard. Largely as a result of this decision, freight traffic failed to materialise, and the railway company chose instead to develop the tourism potential of the area.

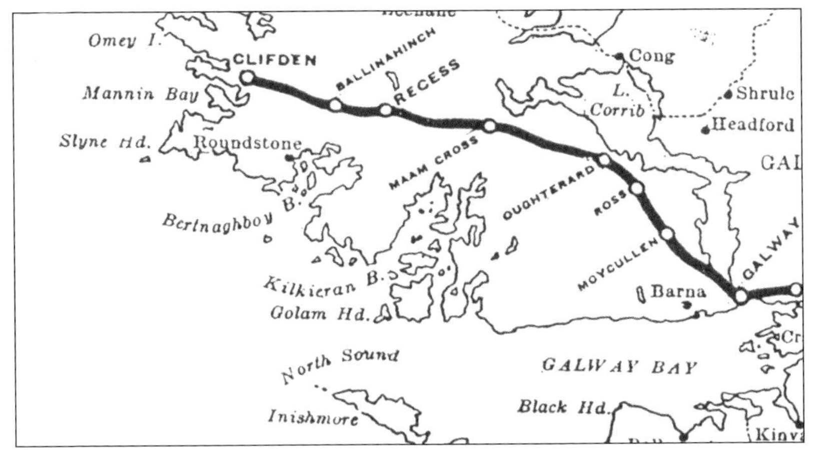

Route of the Galway–Clifden Railway

The total length of the single-track line was just over 48 miles and was built to the Irish standard gauge of 5ft 3in. The line was opened to Clifden on 1 July 1895 and cost £432,000, or £9,000 per mile. Altogether, there were some thirty bridges, including an imposing steel viaduct, which crossed the River Corrib in Galway. Today only the piers remain of the viaduct, the three spans of which were 150ft, with a bascule-type lifting navigation span of 21ft. The engineers for the railway were John Henry Ryan and Edward Townsend. The line was closed in April 1935. (HEW 3256)

C10. Newport Rail Viaduct (L 983939). The Light Railways (Ireland) Act of 1889 specifically targeted districts where the development of fishing or other industries would benefit and where circumstances of distress in those districts necessitated state aid. Under the Act, the Midland Great Western Railway Company (MGWR) constructed branches to Clifden in County Galway and to Achill and Killala in County Mayo. The town of Newport is on the Newport River, where it flows into the northeast corner of Clew Bay. In late 1890, Robert Worthington, the main contractor to the MGWR, began work on the line between Westport and Mallaranny (later extended to Achill). Design was by the railway company's engineer, William Barrington.

Newport road and rail viaducts

At Newport the single-line track crossed the river on a fine viaduct of red sandstone. It has seven segmental arches of 37ft span with a rise of 12ft 6in, the arch rings being 21in thick. The overall length of the viaduct is 305ft and the width 18ft 6in. The piers are 6ft thick with pointed cutwaters. The date 1892 is carved on the parapet, although the line was not opened until 1894 following completion of a nearby tunnel. The viaduct is now maintained as a pedestrian route and, together with the adjacent road bridge, forms an attractive backdrop to the town, especially when floodlit at night. (HEW 3107)

C11. Eglinton and Cong Canals (M 296247 to M 293256 and M 149553 to M 126595). Lough Corrib in County Galway was used from earliest times as a trading route. Improved access to the city of Galway, and even to the sea, had long been the ambition of promoters of various navigation schemes. As early as 1178, the friars of Claregalway Abbey had dug an artificial cut through an island in the River Corrib to shorten the distance to the city. This was later widened and became the main navigation channel.

A basin for shipping at the Claddagh was completed in 1830 to a plan of Alexander Nimmo and John Killaly, but it was not until 1848 that construction of the Eglinton Canal linking Lough Corrib with the sea began. Designed by John McMahon, it was built by direct labour with a grant from the Board of Works. The canal, which could accommodate vessels up to 125ft in length and of 20ft beam, is about three-quarters of a mile long, with two locks, and was opened in 1852. Low bridges were built across the canal in the 1950s, thereby obstructing the navigation.

Work also began in 1848 on the Cong Canal to link Lough Corrib with Lough Mask to the north, but swallow holes in the limestone bedrock, and the decision by the Board of Works in the 1850s to limit the works to the regulation of drainage, resulted in the works being abandoned. (HEW 3090)

Sea lock at Claddagh, Galway

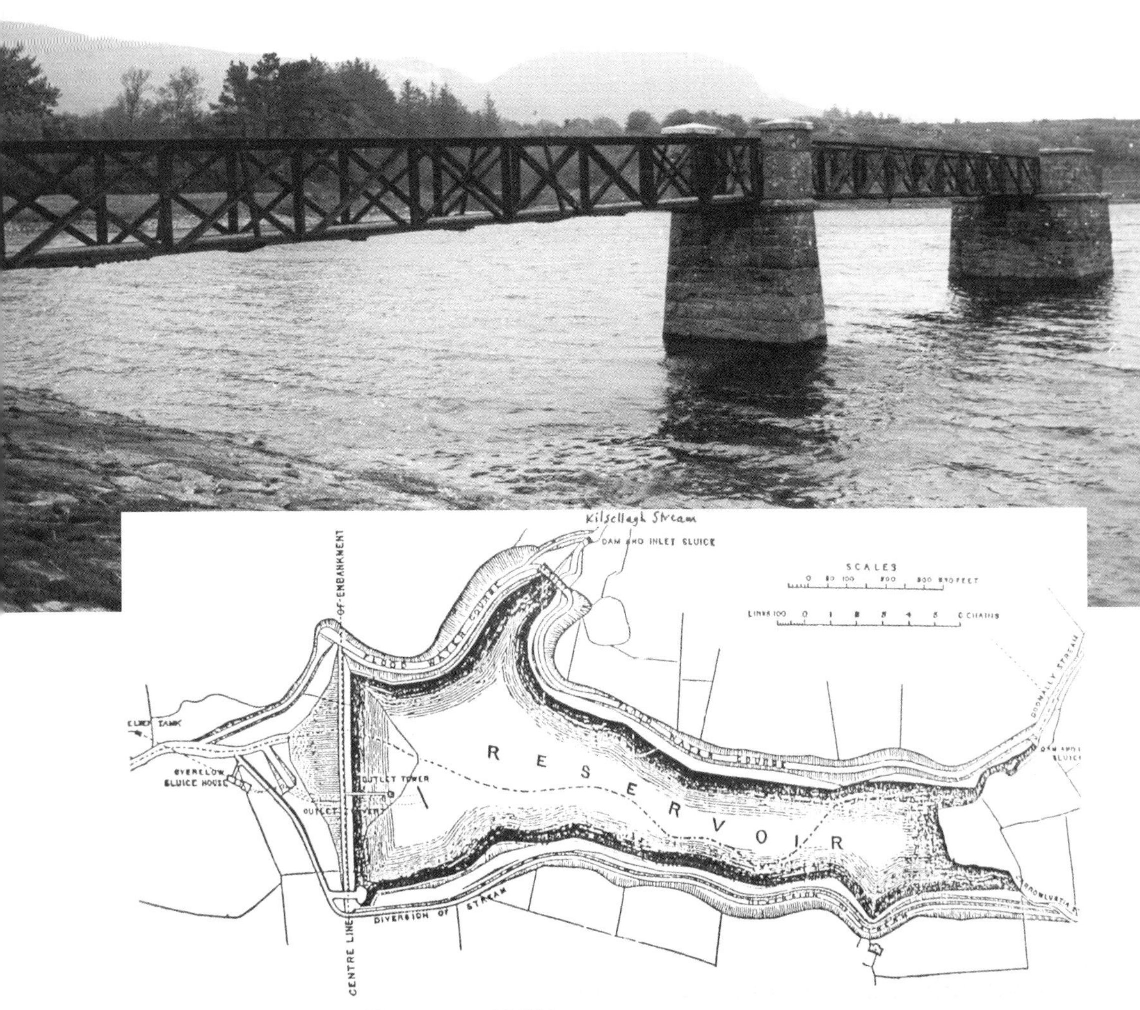

Doonally Reservoir near Sligo, opened 1884

C12. Doonally Reservoir (G 727395). The water supply to Sligo town and its environs comes mainly from Lough Gill to the southeast, but is supplemented by a supply from an earlier reservoir to the northeast of the town. The latter, the Doonally reservoir, part of the Kilsellagh Water Supply Scheme, is a typical example of late Victorian water engineering. The water from the Doonally and Kilsellagh streams is impounded by an earthen dam approximately 30ft high and 600ft long and is led by piped aqueduct to the vicinity of Sligo. A third stream, the Carrowlustia, is led around the south side of the reservoir in a lined channel. The other streams can be diverted around the reservoir in times of flood by similar relief channels on the north side. The inner face of the dam, which has a puddled clay core to prevent seepage, is rubble faced at a slope of 1 in 3, whilst the outer face is grassed. At the southern end of the dam, a semicircular weir, 60ft long, leads overflow water by a 9ft-wide bye-wash, with a segmental-shaped invert, into the Doonally Stream below the dam. The draw-off tower has three valve-controlled openings,

Hyde Bridge, Sligo

and is accessed from the roadway on top of the dam by a wrought-iron latticed girder footbridge of two approximately 50ft spans. The associated water treatment facilities have recently been upgraded in line with EU directives. The Kilsallagh scheme, completed in 1884, was designed by the Westminster consulting engineering firm of R. Hassard and A.W. Tyrell and the contractors were Sweeny and McLarnon. (HEW 3008)

C13. Hyde Bridge, Sligo (G 689364). Sligo occupies a site that has been of importance from earliest times. The River Garavogue, flowing out from Lough Gill to the east, narrows and falls over a series of rock ledges as it nears its estuary. The position of a strategic ford is marked by the older of two town bridges, the so-named New Bridge, a seventeenth-century multi-span rubble masonry bridge with semicircular arches. The second oldest of Sligo's bridges was erected between 1846 and 1852 on the site of an earlier twelfth-century timber bridge. This bridge was itself replaced in the fourteenth century by a narrow eight-arch skewed masonry bridge. It was rebuilt many times before being demolished in 1846 to make way for the present bridge. This is a five-span segmental arch bridge of ashlar limestone, with solid parapets and rounded cutwaters on the upstream face. The spans are

Belmullet Ship Canal

all 21ft 3in with a rise of 6ft 10in, resulting in a level parapet and roadway. Architectural detailing includes a projecting cornice with dentils and archivolts at the top of the arch rings. The piers and abutments are founded on a rock ledge over which the river flows into the tidal estuary. The design was the work jointly of Sir John Benson and the county surveyor of Sligo, Noblett St Leger. The bridge was opened in 1852 as Victoria Bridge, but was later renamed in honour of Ireland's first president, Douglas Hyde. (HEW 3009)

C14. Belmullet Ship Canal (F 702325). To obviate the need for small vessels to be exposed to the full force of the Atlantic west of the remote Mullet peninsula in County Mayo, a short lockless tidal ship canal was constructed by the Board of Public Works in 1855 between Broad Haven and Blacksod Bay near Belmullet. The canal is about 75ft wide across the top and generally about 54ft wide at low water. It is lined with granite and spanned by a later road bridge that limits headroom to around 16ft at high water. Roads in the area built by Nimmo, Bald and others made Belmullet a place of some commercial importance in the nineteenth century. (HEW 3196)

Clapper Bridge near Louisburgh

C15. Louisburgh Clapper Bridge (L 754759). The word 'clapper' comes from the Latin *claperium*, meaning a pile of stones. Such bridges were constructed for pedestrians across rivers at places were the water depth was shallow at times of normal flow and were often associated with an adjacent ford.

The largest complete clapper bridge in Ireland is that located at Burlehinch to the west of Louisburgh in County Mayo. It is around 164ft long and carries a footpath across streams flowing into Roonagh Lough. Agricultural traffic on the associated minor road fords the streams alongside the bridge. The bridge has over thirty spans, the clappers or flat limestone slabs varying in length from 2ft 6in to 5ft. The slabs are generally about 2ft wide and rest on random rubble piers of average height 2ft above bed level. There are 2ft-high parapet walls on the downstream side with 12in by 12in openings centred over each span. These openings allow the water to flow unimpeded during times of flood. No records have been found relating to the construction of the Louisburgh bridge, but local historians believe it to be of 'great age', and it could date from medieval times. (HEW 3241)

ULSTER

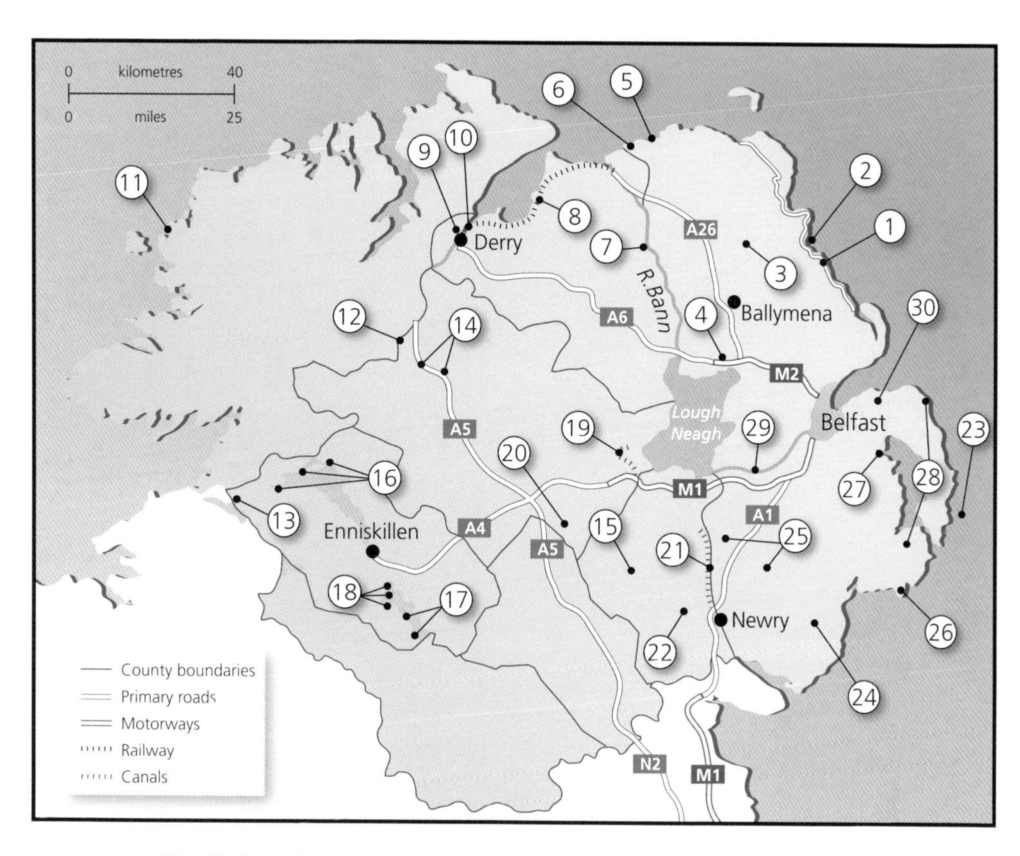

U1 The Antrim Coast Road
U2 Carnlough Harbour
U3 Dungonnell Dam
U4 Randalstown Bridges
U5 Giant's Causeway Tramway
U6 Portrush Harbour
U7 Lower Bann Drainage and Navigation
U8 Coleraine–Derry Railway
U9 Craigavon Bridge, Derry
U10 Foyle Bridge
U11 Cruit Island Footbridge
U12 Clady Bridge
U13 Erne Drainage and Hydroelectric Development
U14 Newtownstewart and Ardstraw Bridges
U15 Tassagh Viaduct
U16 Boa Island and Rosscor Bridges
U17 Trasnagh Island Bridges
U18 Upper Lough Erne Bridges
U19 Ducart's Canal
U20 Ulster's Dredge Bridges
U21 Newry Canal
U22 Craigmore Viaduct
U23 South Rock (Kilwarlin) Lighthouse
U24 Silent Valley and Ben Crom Reservoirs
U25 White and Seafin Bridges
U26 Killough Windmill
U27 Nendrum Tide Mill
U28 Hunt's Park, Donaghadee and Downpatrick Water Towers
U29 The Lagan Navigation (Sprucefield–Lough Neagh)
U30 Crawfordsburn Viaduct

Above: Antrim Coast Road – Glendun Viaduct

Below: Carnlough Harbour: 'Loading' wall and quay

U1. The Antrim Coast Road (D 155407 to D 407035) was one of the first major roads built under the provisions of the Public Works (Ireland) Act of 1831, and was 33 miles long. The road was to open up the coastal towns previously served by steep mountainous roads, largely impassable in winter. The design, prepared by the Board of Works engineer William Bald, was undertaken in sections. Where it runs beside the sea, the road was constructed by blasting rock from the overhanging cliffs onto the foreshore and building on the resulting berm. This road became the main route between coastal towns, but these sections have constantly given trouble from wave and frost action, together with small areas of mud slips and rock falls, usually in wintertime. Finally in 1967 two very major collapses of rock occurred which closed the road to traffic completely. Antrim County Council set about rebuilding the route themselves, which was not without its difficulties. Design and construction is reported in a paper presented to the NI Association of the Institution of Civil Engineers in 1972.

The road rises steeply from Cushendall with the three-span Glendun Viaduct (HEW 2427) forming the elbow of a hairpin bend. Although the overall concept of the road was undoubtedly Bald's, the detailed design for this viaduct is widely attributed to Charles Lanyon, when he was county surveyor of Antrim. The central arch spans over the River Dun, the northerly arch over a minor road, with the third arch standing on a wooded slope (D 215322). Most of the other smaller stone bridges on the road still remain.

The section of the road to the north of the Glendun Viaduct (D 225324 to D 230326) was constructed at right angles to a 35° ground slope, and is, in parts, held by 'dry stone' retaining walls up to 15ft high, with inward piers, such that the road ballast becomes integral with the walls (HEW 2428).

North of the viaduct the road rises over Cushleake Mountain which was a magnificent feat of road construction. Bald undertook the progressive drainage of deep peat to allow work to proceed. Today the total road is one of the region's major tourist attractions, although it still suffers from the occasional problem in winter. (HEW 1067)

U2. Carnlough Harbour (D 288100) lies on the east coast of County Antrim, north of Larne. An early pier of 200ft by 20ft was in existence around 1800. The Londonderry family inherited lands at Carnlough in 1834 and improved the harbour to facilitate the export of lime to Scotland for use in iron smelting. The limestone was ferried in small lighters out to ships at anchor in the bay, which is not particularly well sheltered from westerly winds. Work began in 1853 to construct a sheltered basin. Unfortunately, as the excavation proceeded seaward, problems were encountered when an igneous dyke was struck. It was not until late in 1855 that the basin became operational, and then only for ships with a maximum draught of 8ft.

The original pier of 1800, retained and strengthened for protection, cut right across the longshore drift movement and siltation began at once. In 1859 a dredger removed a large amount of material and the rest of the dyke, but only when part of the south pier collapsed and was rebuilt in 1860 did the basin prove to be a moderate success, providing dredging was maintained.

The three-quarters of a mile of inclined railway, built to the English gauge of 4ft 8½in, runs up to the quarry and was operated by horse (and gravity) until 1898, steam until 1952, and electric traction until its closure. A branch line over a mile long with a gauge of 3ft 6in operated from 1890 until 1922. The inclined plane required a 200yd cutting through rock and earth with bridges across the back and main streets built in white limestone, the track eventually terminating at chutes over the vessels at the quayside. Ammonium sulphate (extracted from peat) was exported from 1902 to 1908, and processed lime from 1855 until closure in 1965.

U3. Dungonnell Dam (D 192171). Completed in 1970, Dungonnell Dam was the first dam built in either Britain or Ireland which was waterproofed using an asphaltic lining laid on the upstream face. Six such dams had been built previously elsewhere in Africa and Norway. The dam itself is of straightforward design and construction with regard to cut-off and overflow arrangements and the resulting reservoir provides a water supply for Ballymena. The dam is built of general rockfill laid to slopes of 1 in 1.5 downstream and 1 in 1.7 upstream. It is some 950ft long with a maximum height of 52ft. It has a concrete cut-off wall and grout curtain near to the front toe. The upstream face consists of layers – 6in of small stone, 3in of asphalt concrete underseal, and 5in of open bitmac drainage layer, all sealed by 4in of asphaltic concrete. Although the contractor had some problems in the design of equipment to lay the waterproofing to the required slope, these were overcome and the work proceeded without difficulty. The main contractor was John Rainey (Construction), Portrush, with the waterproofing by Wimpey Asphalt of London and the overall design by Ferguson and McIlveen of Belfast. (HEW 1854)

Dungonnell Dam

Randalstown Rail Viaduct

U4. Randalstown Bridges (J 084902). At Randalstown, County Antrim, is a striking railway viaduct in cut stone. This was completed in 1855–56 as part of the extension of the branch railway line to Randalstown from the Belfast and Ballymena Railway. The viaduct has basalt stonework piers and facings to its eight arches, the barrels themselves being completed in brick. The track was single, although with an overall width of 31ft 4in the structure appears to be wide enough for a double track, similar to a number of structures on this line. The eastern approach crossed the Belfast road by a girder bridge, which has been removed since the line closed in 1959. The first arch crosses a (now disused) mill race, the next six cross the River Main and the last is over a bankside road. Spans are each approximately 30ft and the height of the parapet over the riverbed is 52ft. The design was by Charles Lanyon and the contractor was William Dargan. In the shadow of the viaduct is one of Ulster's fine, virtually undateable, multi-arch stone road bridges. This is at an early crossing point on the River Main and the road here was one of the Irish turnpikes. It is stated in the Ordnance Survey Memoir for Randalstown that this bridge is a 'clumsy old structure of some antiquity'. It was twice widened, lastly in 1816, and consists of nine semicircular arches of unequal span and is 193ft long by 29ft wide. (HEW 2130)

Giant's Causeway Railway

U5. Giant's Causeway Tramway (C 858404 to C 035411). In 1881, work started on the construction of a 3ft-gauge tramway to connect Portrush railway station with the Giant's Causeway, via the nearby town of Bushmills. The engineer in charge was William Traill. The track ran alongside the road. The only structure of any note was a through lattice-girder three-span bridge over the River Bush. The feature which makes the tramway important historically was the decision to run the trains by electricity, to be generated by a hydroelectric plant located in the River Bush upstream of Bushmills, which was an early use of electric traction and which is claimed to be the first using hydroelectricity. The line opened in 1887. After problems with electrical leakage to earth Traill devised an overhead pick-up system. The line closed as being uneconomical in 1949.

In 2001 the line was re-laid from Bushmills to the Causeway using track, a steam train, and coaches brought from Shane's Castle, near Randalstown. The bridge over the Bush was replaced with a similar, but lighter, structure manufactured from hollow steel sections. At Bushmills, a small wooden halt was built and, at the Giant's Causeway, new buildings, including an engine house, were built. (HEW 0573).

Portrush Harbour

U6. Portrush Harbour (C 855407). Portrush is recorded as a landing place from the sixteenth century but the main harbour in the area was at Colcraine. However, access was difficult over the bar of the River Bann. In 1803 John Rennie reported to the merchants of Coleraine regarding a possible harbour at Portrush. He proposed a harbour 150ft long for a cost of about £11,000. His report was accepted, the work was carried out, and completed around 1820 (this is now the inner harbour which contains much of the original stone work). The harbour soon proved to be too small as it was only 0.5 acre at half tide and 0.25 acre at low water. Sir John Rennie (his son) recommended his favourite design consisting of two arms built out from the shore and curving in to leave an entrance. This new harbour comprised a north pier running southwest from Ramore Head for 579ft and a south pier of 715ft running west-southwest and turning north. This gave an entrance of 270ft and encompassed 10 acres. In 1835, the works were almost complete. The south pier comprises random rubble but near the outer end the inside face has been strengthened with concrete blocks and a concrete top slab dated 1894, the better to resist Atlantic gales which break over the piers. The north pier has a fine outer masonry apron at a slope of 2.4 to1. There is a short concrete extension (1887) carrying pilot lights. Virtually all of the original works of John Rennie and his son, Sir John, remain visible. (HEW 2647).

U7. Lower Bann Drainage and Navigation (H 988905 to C 856303). Lough Neagh is the largest lake in either Britain or Ireland. Several rivers and streams flow into the lough, but only the Lower Bann River discharges from it. The total catchment is 2,216 sq. miles. For over two centuries, the flooding caused by Lough Neagh led to repeated calls for remedial action.

The first major flood relief works were undertaken in 1840 under the direction of John McMahon, Chief Engineer of the Board of Works, and cost over £250,000. Half of the cost was for the relief of flooding, the other half for making the river navigable from Toome to Coleraine. There were six steps in water level (at Toome, Portna (two steps), Movanagher, Carnroe and the Cutts at Coleraine). Lock chambers were 120ft by 20ft. Fixed weirs were positioned at Movanagher and Carnroe, locks and associated short canals exist at all five locations to facilitate shipping. These works improved the drainage considerably but by 1859, it was being alleged that the expected relief from flooding was not occurring. No further works were undertaken and it was not until 1925 that Major Shepherd, Chief Engineer of the Ministry of Finance for Northern Ireland, arranged for new surveys to be carried out.

Shepherd found that McMahon's scheme had not been completed. The proposed minimum cross-section of 2,400sq.ft was found in numerous places to be of less than 900sq.ft. Shepherd concluded that, as McMahon had died before the completion of the scheme, his successors, faced with increasing costs, had reduced its extent. McMahon had calculated that the weir at Toome would discharge 400,000cu.ft/min. when Lough Neagh rose to 53ft ordnance datum (OD). However, with the restricted channel, the Toome weir

Lower Bann at Carnroe

was drowned at 52.25ft OD. In 1930, an order was placed with Ransomes & Rapier of Ipswich for new sluice gates (five at Toome Weir, four with fish passes at Portna and four at the Cutts, all 60ft wide by 9ft 6in high). The work was undertaken by James Dredging Towage and Transport of Southampton, who sublet the rock excavation and weir and sluice work to Walter Scott & Middleton of London. Since the completion of Shepherd's scheme, which established a water level of 54ft OD, there have been two further lowerings of the lough, each time achieved by adjustment of the sluice gates. The level is now 50ft OD. The three sluice gates at Toome, Portna and the Cutts have recently been refurbished and the operations electrified, so that the original 1930s construction is still operating as installed. (HEW 2122)

U8. Coleraine–Derry Railway (C 440164 to C 844326). This railway was incorporated in August 1845, the inception and finance coming mainly from London businessmen. Work was started on two short tunnels, 668yd (J 758363) and 307yd (J 761363) long, between Downhill and Castlerock – the first tunnels in Ireland to be driven using explosives. Subsequently, on 8 June 1846, 30,000 tons of rock was blasted from the cliffs west of the tunnels to form the base for the track. The promoters had intended to build the single-line track on an embankment carried across the slob of Lough Foyle, thereby reclaiming 22,000 acres of land. However, this proved to be more difficult and costly than anticipated and the route had to be moved landwards, although 10,000 acres were reclaimed. The route across the lough necessitated the building of two long viaducts over the estuaries of the rivers Faughan and Roe. Initially these were in timber but were prone to damage from hot ash from the locomotives' fire-boxes. Under William Wallace these viaducts were replaced by concrete T-beams in spans varying from 29ft to 32ft 6ins. The Faughan Viaduct (J 490225) was replaced in 1937 in eleven spans and the Roe Viaduct (J 643294) in 1938 in fifteen spans. The line from Derry to Limavady was opened in 1852 and through to Coleraine the following year. The system was isolated until the first (timber) bridge over the River Bann at Coleraine was completed in 1860 which allowed through running to Belfast. This bridge was replaced by eleven 70ft spans of steel by Wallace in 1924. To allow the passage of river traffic an opening bascule span was included which incorporated the Strauss method of counterbalancing the steelwork of the lifting span. (HEW 0700).

Roe Viaduct on the Coleraine Derry railway

U9. Craigavon Bridge, Derry (C 437161). The first bridge across the River Foyle at Derry was constructed as a toll bridge in 1791. In 1863 it was superseded by the Carlisle Bridge, which was also a toll bridge until New Year's Day, 1878. The upper deck carried road traffic and the lower deck the mixed-gauge railway lines of 5ft 3in and 3ft which linked the standard and narrow gauge stations on either side of the river. It was 30ft wide, built of iron, and cost £90,000. In 1933, the Carlisle Bridge was replaced by the Craigavon Bridge, which has a wider carriageway on the upper deck – 40ft plus 10ft-wide footways. It again accommodated the mixed-gauge railway goods lines on the lower deck. At either side of the river there was a turntable for the wagons, which were hauled across by ropes operated by capstans. The bridge has five main spans, each of 130ft, and seventeen approach spans, giving a total length of 1,260ft. To find a firm foundation, cylindrical piers were sunk to a maximum depth of 69ft below HWOS tides. The main Pratt truss type girders are spaced 36ft apart and incorporate 5,000 tons of steel. The Northern Ireland government provided £251,000 of the total cost of £255,500, leaving Derry Corporation to pay the balance. The bridge was designed by Mott, Hay and Anderson and built by Dorman Long & Co. Work commenced in 1931 and the bridge was opened on 18 July 1933. In 1968, after the closure of the Great Northern Railway track on Foyle Road in 1965, the lower deck was converted to a two-lane roadway. In 1984, the traffic congestion on the Craigavon Bridge was relieved by the opening of the Foyle Bridge some two miles downstream. (HEW 2127)

Craigavon Bridge, Derry

Foyle Bridge ESLER CRAWFORD PHOTOGRAPHY

U10. The Foyle Bridge (C 436178) near Derry has the largest single span of any bridge in Ireland. It was built to relieve congestion on the nearby Craigavon Bridge, and was opened in October 1984. Design and build tenders were sought in 1977 and consultants Ove Arup & Partners were appointed to assess the five bids received. In December 1979, a tender for £15.7 million was accepted. The contract was awarded to a joint venture partnership between Redpath Dorman Long and Graham of Dromore with Freeman, Fox & Partners as their designers. After a twelve-month design contract, construction began in December 1980. The winning design consisted of two parallel steel box girders with curved concrete box-section approach viaducts. The total length of the crossing is 865.7m. The main bridge has side spans of 144.3m and a central span of 233.6m. The substructure consists of twin concrete piers on spread footings on rock. Each steel box girder was built in three parts by Harland and Wolff, Belfast, and transported to the site by barge. The four side-span units, each 180m long and weighing 900 tonnes each, were first placed. The two central-span units, weighing 700 tonnes each, were then lifted into position and bolted before final welding. Navigation requirement was for a clearance over MHWS tides of 32m over a width of 130m. Prestressing of the concrete spans was on the PSC Freyssinet K system, using multi-strand tendons. The bridge, the profile of which was approved by the Royal Fine Art Commission, accommodates dual 7.3m carriageways and 1.8m footways and a central reservation. (HEW 2131)

Cruit Island Footbridge

U11. Cruit Island Footbridge (B 730186). This reinforced concrete footbridge, which connects Cruit Island to the mainland in County Donegal, was built for the Congested Districts Board in 1911. It was designed by Mouchel & Partners and constructed using the Hennebique system of reinforcement.

The bridge has an arch slab 10in thick, being 7ft wide at the springing and narrowing to 4ft 9in at the crown. The rise is 17ft and the span some 90ft. The deck spans about 110ft between masonry abutments (which previously carried a timber deck) and is supported off the arch by three vertical slabs on each side.

This is an interesting early example of reinforced concrete bridge work and it is a pity that at the time of writing it is out of use and is being allowed to decay. (HEW 3150)

U12. Clady Bridge (H 294940) has seven arches spanning the River Finn – the border between County Tyrone and County Donegal. It is estimated to have been constructed before 1714 on the site of an important ford.

Clady Bridge has all of the characteristics of medieval construction. There are nine arches, each with a different size and a near semicircular profile. The piers are wide with triangular cutwaters which are carried up to the parapet level forming deep refuges. The road is very narrow (12ft wide) and controlled by traffic lights. The border between the two jurisdictions is clearly marked on the parapet over the central span. (HEW 2133)

Clady Bridge

U13. Erne Drainage and Hydroelectric Development (G 873612 to G 930605). The 59-mile long River Erne, the second largest river in Ireland, drains 1,525 sq. miles of southwest Ulster. The river flows through Upper and Lower Lough Erne, discharging at some 164ft above mean sea level at Belleek, and then dropping quickly by a series of rapids to the sea. The catchment regularly flooded because of the restricted discharge of the channel at Belleek, where the river runs on rock.

In 1863, the Lough and River Erne Drainage District Board was set up and constructed four flow-control sluices at Belleek. The first drainage scheme undertaken by the Board was in 1885–90, but was restricted by an inability to excavate the rock at Belleek.

In 1918, the Board of Trade established the Water Power Resources Committee. Under the chairmanship of Sir John Purser Griffith, this body reported in 1921 on the potential for power generation in Ireland. In the late 1920s, work commenced on the Shannon Scheme, and it was not until 1945 that an Act was passed by the Irish government to allow work to commence at Ballyshannon. A corresponding Act was passed by the Northern Ireland government in 1947.

The works involved were the Cathaleen's Fall Dam, 844ft long, 90ft high, with an installed capacity of 45MW; the Cliff Dam, 690ft long, 60ft high, with an installed capacity of 20MW; the improvement of 3½ miles of river channel at Belleek, which involved the removal of 800,000 cu.yd of material, mainly rock, under a target price contract, and was undertaken at a cost considerably less than the £1,125,000 estimate; and the removal of the Belleek sluices and the construction of the Portora barrage at Enniskillen for water-level control in Upper Lough Erne. Construction work was undertaken by the Cementation Company Ltd and was completed by 1956 (HEW 3243).

Erne Hydroelectric Scheme
ESB ARCHIVES

U14. Newtownstewart and Ardstraw Bridges are built in rubble-stone work and date from the early eighteenth century. In both cases, the arches are semicircular with triangular cutwaters which are carried to the soffit level of the arches.

The bridge at Newtownstewart (H 403858), over the River Strule, has six arches. On Moll's map of 1714 the road is shown as crossing the river here before proceeding on the opposite bank to Strabane (HEW 2140). Today, the bridge has been largely bypassed by the adjacent Abercorn Bridge, a concrete structure designed by Mouchel & Partners and opened in 1932. It is interesting to note that the eighteenth-century masonry bridge survived the catastrophic floods of 1929 whereas the later Abercorn Bridge was swept away and had to be replaced.

At Ardstraw, the four main arches carry the B164 road over the River Derg. However, Moll's map shows the main road to Derry crossed here apparently by a ford. Ardstraw Bridge (H 349873) has been dated to 1727. The arches are approximately 12ft span and the piers are 5ft wide. On one bank there is a wider buttressed pier followed by a further arch providing flood relief. (HEW 2139)

Newtownstewart Bridge

U15. Tassagh Viaduct (H 861383) lies in a remote area some 2½ miles north of Keady in County Armagh, and was constructed as part of one of the last sections of railway line in Ireland. The line connected Armagh, Keady and Castleblayney and was partly funded by the Great Northern Railway Company. The line never proved economical and the southern portion closed after the partition of Ireland. However, the line between Keady and Armagh continued with goods traffic up to 1957, passenger traffic having ended in 1932.

The viaduct with eleven arches was built between 1903 and 1910, constructed with concrete piers and brick vaulting to the arches. The Callan River flows through one of the arches, a second arch spanning a minor road. The deck is asphalted to prevent water ingress to the structure, which is in remarkably good condition, but access is blocked by heavy locked gates. (HEW 1975)

Tassagh Bridge, parapet details

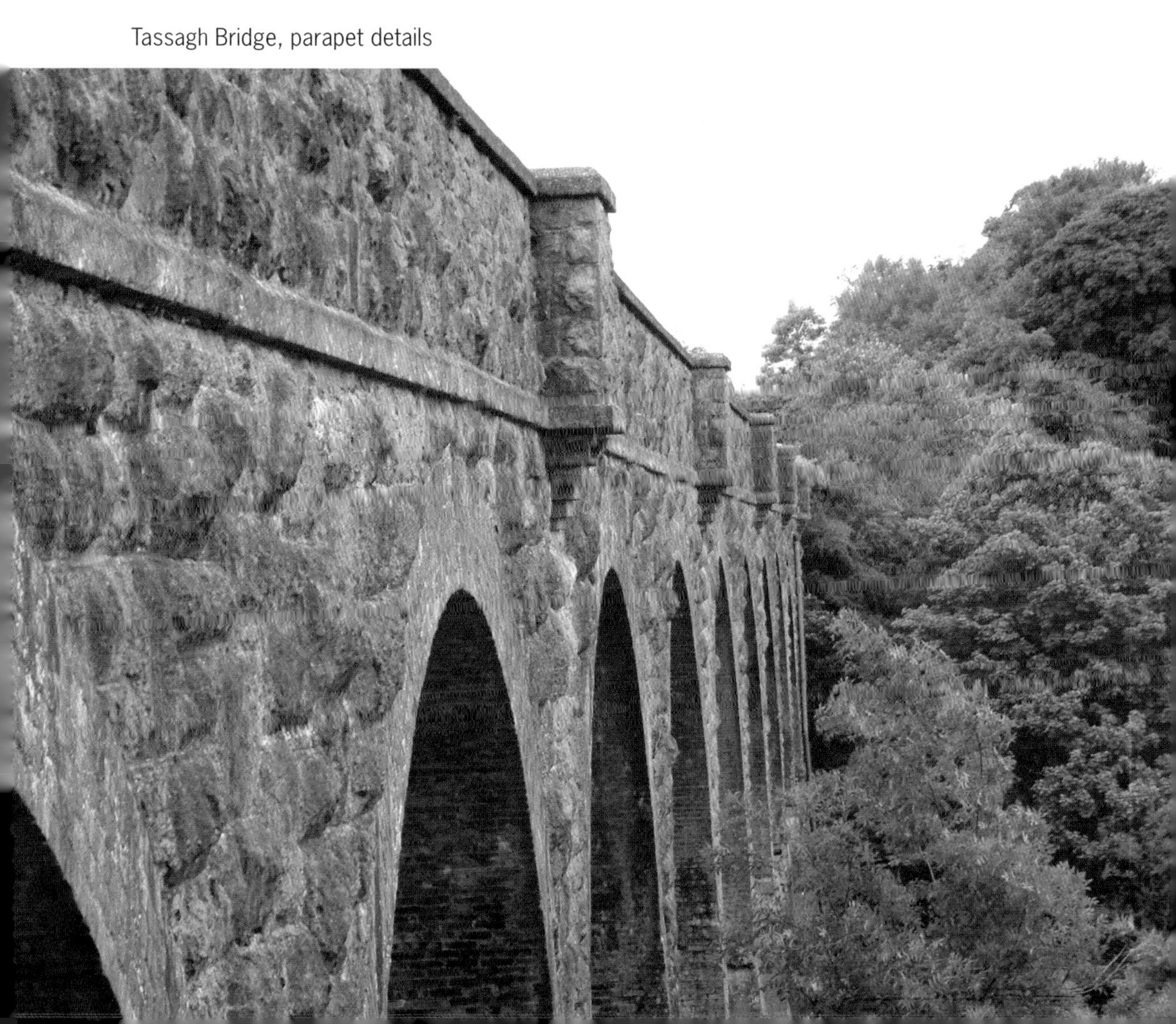

U16. Boa Island and Rosscor Bridges. The partition of Ireland in 1921 affected the road pattern in County Fermanagh and led directly to the construction of a series of multi-span concrete bridges. Boa Island, 5 miles long, lies on the north shore of Lower Lough Erne, which up until 1921 relied on ferry services to the mainland. Given its geographical relationship to the new border and acknowledging the islanders' difficulties with transport, the new Northern Ireland government agreed to construct two new bridges, one at each end of the island, replacing the ferry connections and keeping the road within Northern Ireland. The number of spans was limited to three by the construction of long approach embankments. The Portinode Bridge (H 138644), at the eastern end of the island, has a 700ft-long embankment, whilst the western Inishkeeragh Bridge (H 064625), has a 1,000ft embankment. A third bridge was constructed 6 miles to the west, connecting roads on the north and south sides of the lough, keeping travel within Northern Ireland. Rosscor Viaduct (H 987586) has nine spans crossing the River Erne and has been designed and constructed in a similar manner to the Boa Island bridges, all three by the Trussed Steel Concrete Company.

The three bridges are supported on *in-situ* concrete slab walls carried on precast concrete piles without capping beams, the outer piles raking away from the bridge, visible above water level. Rosscor Viaduct has a breakwater enlargement cast into the lower part of the vertical walls on some of its central spans. The spans are 35ft wide, and the road width is 16ft, with concrete parapets. (HEW 2123)

Rosscor was damaged by an IRA explosion in the 1970s, and a Bailey bridge still remains in position over the first two spans. In the photograph, the fallen spans and the Bailey bridge can be seen at the far end.

Rosscor Bridge

Lady Craigavon Bridge, Trasnagh Island

U17. Trasnagh Island Bridges. Following partition in 1921, there was a need to provide a route from the west side of Upper Lough Erne, which would avoid travelling across a state border (with customs posts), or suffer a long detour. It was decided to use a route across the lough at Trasnagh Island, which was served by chain ferries, replacing them with multi-span road viaducts constructed in reinforced concrete. Both bridges were designed by L. G. Mouchel & Partners Ltd., and built in 1934 by A.E. Farr Ltd, London, using the Hennebique reinforcement system. The bridges are on the B127 road, between Derrylin and Lisnaskea.

The Lady Craigavon Bridge (H 330278), on the east side of the island, has nine main spans of varying lengths totalling 410ft, a 16ft-wide roadway with 3ft 6in high concrete parapets. The shorter side spans are supported by pairs of square concrete columns braced with cross walls and carried on precast single piles just visible below each column. The wider central navigation spans are supported by larger concrete circular columns with cross walls carried on multiple piles. The main 8½in concrete deck slab is carried on cross beams at 5ft centres supported on each side by a deep edge beam. Pedestrian refuges are built into the parapets over each central circular column.

The Lady Brooke Bridge (H 320273), on the west side of the island, has thirteen spans, all in a similar construction to the side spans of the Lady Craigavon Bridge, with a total length of 510ft. (HEW 2124)

U18. Upper Lough Erne Bridges. There are three bridges connecting the islands of Inishmore and Cleenish, lying at the north end of the lough.

Inishmore Viaduct (H 254342) connects the island to the west shore of Upper Lough Erne, one of very few metal road viaducts in Ulster. The pavement is carried on steel arched plates, spanning between longitudinal side girders that also form the parapets. The central span of five has a bow girder on each side braced together on top (this bracing was raised later to meet modern headroom requirements). Four heavy masonry piers on concrete plinths support the bridge. It was designed by James Price and built by direct labour. Construction commenced in 1891, but difficulties were encountered with the embankments, and the bridge was not finished until 1900, by the contractor Andrew Mair. (HEW 2125)

Carry Bridge (H 294373), which connects the island to the east side of the lough, is the most modern of the three bridges, and is the only one constructed in concrete. It has a single 140ft-span thin concrete arch, with a reinforced concrete deck and concrete parapets supported off the arch directly at the centre and by four portal frames supported off the arch on each side, giving a very light and open appearance. The bridge was designed by Ministry of Finance (Northern Ireland) and was opened in 1956. (HEW 2645)

Further north, at H 256385), the road to Cleenish Island is carried on a three-span steel Callender-Hamilton Type B (heavy-duty road bridge) constructed by British Insulated Callender's Construction Co. Ltd in 1953. Fermanagh County Council built the abutments. This type of structure comprises Warren girders, supplied in multiples of 10ft and using commercially available steel sections. The only non-standard parts appear to be the gusset plate connectors and end bearing assemblies. First described in 1935, large numbers of this sort of bridge were built during and after the Second World War to strengthen masonry bridges considered too weak for military loads, or to replace damaged bridges. Structures of this sort are usually considered as temporary, but there are no plans to replace this one; indeed a concrete deck was added in 1986. (HEW 1972)

U19. Ducart's Canal (H 804645 to H 844665). The mining of coal at Coalisland in County Tyrone was a long-established tradition, although the seams show severe faulting. Transportation of the coal was a problem, but was relieved by the opening of the Newry Canal in 1742. In 1753, proposals were made for an extension of the as yet uncompleted Coalisland Canal to Drumglass. This extension was completed in 1777 and was known as Ducart's Canal. The method of working of the extension was unique in Ireland. Ducart was a Sardinian architect (Daviso de Arcort), whose building works in Ireland include the Custom House (now the Hunt Museum) in Limerick. The coal was conveyed in small wheeled tubs, which were run on to lighters on the canal, which consisted of three level sections connected together and to the existing canal by stone ramps, known as 'dry hurries', down which the tubs were run to the next level. Operation was more difficult than

Above: Inishmore Viaduct

Below: Carry Bridge

expected and Ducart's Canal soon fell into disuse. Parts of the hurries remain, the one most easily seen being at Furlough beside the Dungannon–Newmills road at H 812665. The most prominent work still extant is the ashlar stonework aqueduct that carried the canal over the river Torrent at Newmills (H 815673). This has three near semi-circular arches and remains in an unaltered state, although it is now a viaduct (HEW 0445)

U20. Ulster's Dredge Bridges. Ulster originally had several bridges built to the patent of James Dredge of Bath. The Dredge patent was innovative for its time using wrought-iron suspension chains, which increased from a single bar at mid-span by adding a further bar at each link position back to the support. Thus the longer the span, the more elements there were in the cable. This arrangement had the advantages of less weight and a shorter erection time. The suspension chains pass over the cast-iron support to a ground anchor. The elements are connected together with bolts and shear pins.

At Glenarb (H 762447) near Caledon is a Dredge patent suspension bridge which was cast by the Armagh Foundry and opened originally in 1844. The main span is about 78ft with side spans of 20ft, and a width of 3ft 6in, which is unusually narrow for a Dredge bridge and exhibits a more elaborate bracing system than that usually seen, with noticeable

Newmills Aqueduct

movement underfoot. The bridge's original site inside Lord Caledon's estate was affected by a River Blackwater drainage scheme, so in 1984 the bridge was dismantled, refurbished, and re-erected in 1990 on new foundations over the same river, in a small park just outside the estate. (HEW 1857)

Lord Caledon's estate contains a second Dredge bridge (H 745432), which has a single 73ft span on masonry abutments and a deck width of 10ft 6in. It was also cast by the Armagh Foundry in 1845. This bridge is still in use for light agricultural vehicles and is well maintained by the estate. (HEW 1856)

At Moyola Park, Castledawson (H 927936), there were originally two Dredge footbridges. One of these, built about 1846, was swept away in 1929. The remaining example has two 66ft spans either side of a single masonry pier in the river. It dates from 1847 and has a width of 5ft 9in. It is in poor order, but attempts are being made to restore this important grade 'A' listed structure. (HEW 1858)

The only Dredge bridge which was not a footbridge – that at Ballievy, near Banbridge (J 158451) –was totally destroyed in 1988 when a large lorry attempted to cross it, even though there was a two-ton weight restriction in force. (HEW 1855)

Dredge bridge in Caledon Park

U21. Newry Canal (J 087250 to J 022526). Coal was discovered in County Tyrone in the seventeenth century and a canal to link the mines with Newry via Lough Neagh and thence Dublin was first proposed in 1703, but no action was taken until 1732 when work began under Edward Lovatt Pearce (Surveyor General), but he died in 1733. Richard Cassels took over until 1736, followed by Thomas Steers. The canal was completed in 1742, which makes the Newry Canal the first canal to be constructed in either Britain or Ireland, the Sankey Brook Canal not being authorised until 1755. The canal was 18½ miles long, rose to 78ft above sea level, was 45ft wide and 5ft to 6ft deep. From Newry, there were eleven ascending locks and three descending locks to Lough Neagh. The lock chambers were 44ft by 15½ft and between 12ft and 13½ft deep. Problems arose as early as 1750, when the water supply to the summit from the nearby Lough Shark was found to be inadequate, so water was brought from the Cusher River via a canal and the still extant ten-arch aqueduct (J 053446) at Terryhoogan. In 1800, John Brownrigg reported that the canal was in a ruinous state, including the lock chambers, which had been built of brick faced with Benburb stone and with 2in-thick deal floors. Four locks and three bridges were rebuilt, the summit length widened and access to the Newry Ship Canal (1769) improved. Work was completed in 1811. One of Brownrigg's bridges can be found at Jerrettspass (J 064333).

Newry Canal

The canal now entered a period of prosperity, with tonnage doubling in the 1830s to 100,000 tons per annum. In the 1850s it carried 120,000 tons per annum and yielded £4,000. Railway competition, however, began to reduce this and by 1918 tolls were no longer covering expenditure. By 1939 the canal was derelict, and it was abandoned in 1956 as a navigation, although it still exists as a drainage watercourse. (HEWs 1870, 1900, 1901)

U22. Craigmore Viaduct (J 067284). The railway line between Dublin and Belfast was constructed in sections by three different companies between 1837 and 1855. The Egyptian Arch (one of only a few structures in Ireland built in the Egyptian style, then in vogue) and the impressive Craigmore Viaduct, which opened to traffic on 13 May 1852, were built near Newry. Both were designed by Sir John Macneill and built by William Dargan. The Craigmore Viaduct has eighteen semicircular arches, each of about 59ft 6in span with piers 7ft 6in thick at springing level. The width of the trackway is 28ft. The height of the Craigmore track above of the Bessbrook River is 137ft, making it the tallest railway bridge in Ireland. It was built of local Newry grano-diorite and it cost about £50,000. The station for Newry is located near the southern end of the viaduct. (HEW 1973)

Craigmore Rail Viaduct

U23. South Rock (Kilwarlin) Lighthouse (J 678531) lies off the Ards Peninsula and presents a hazard to coastal shipping. The building of a lighthouse on the South Rock commenced in 1795 to a design by Thomas Rodgers, contractor to HM Revenue Commissioners and took a number of years to complete. It was built of granite blocks, the end joints of the tiers being joggled into each other. Eight iron bars, 3in square, run within the wall, being keyed into circular plates every third or fourth course. The tower is 37ft in diameter at the base, tapering to 17ft under the cornice at a height of some 70ft. The first 22ft is of solid masonry and this design, for a wave-washed lighthouse, was considered to give the greatest strength. It pre-dates Robert Stevenson's taller Bell Rock Lighthouse off the east coast of Scotland (1807–10) and it is reported that he visited Kilwarlin during the construction of Bell Rock.

In 1810, responsibility for Irish lighthouses was vested in a new authority, the Corporation for Preserving and Improving the Port of Dublin (commonly known as the Ballast Board). The lighthouse was built on South Rock only 2 miles off shore, but despite the single revolving white light, wrecks continued to occur. Accordingly, the light had to be abandoned in 1877, and a lightship stationed a further 2 miles out to sea which, in turn, was replaced by a navigation buoy in 2009.

Despite the intervening years, the structure has survived the ravages of tides and storms remarkably well and, although unlit, still acts as a navigation mark for vessels making an inshore passage. (HEW 2098)

U24. Silent Valley and Ben Crom Reservoirs (J 306230 and J 317260). In 1891, the Belfast City & District Water Commissioners asked their Consulting Engineer, L.L. Macassey (1843–1908), to report on further sources of water for Belfast. He reported that it would be necessary to develop a scheme in the Mourne Mountains. Accordingly, the catchments of the Kilkeel and Annalong rivers were taken by an Act of 1893. A 40-mile long pipeline (including two long tunnels in rock) was constructed to deliver water to a reservoir at Knockbreckan (J 365663), near Belfast, flow commencing in 1901. Construction of the associated storage reservoir was deferred until 1923, when S. Pearson & Sons, represented by Sir Ernest Moir, was awarded the contract for the building of the Silent Valley Dam, designed by F.W. McCullough (1860–1927) to store 3,000 million gallons. The dam was to be an earth embankment 1,500ft long and 88ft high, with a clay core and concrete cut-off trench. Site investigation boreholes indicated rock at between 20ft and 75ft. The contractor started to excavate the cut-off trench, but found no rock, only large boulders interlaced with fine saturated silt. Rock was eventually found at 140–200ft below ground level. After arbitration in 1926, the contractor proceeded on a cost-plus basis, using compressed air. A series of cast-iron shafts with a diameter of 11ft 10in was

Facing page: South Rock (Kilwarlin) Lighthouse GORDON MILLINGTON

sunk and these were used alternately to pump and excavate. (An example of the shafting can be seen on site beside the visitors' centre). Construction proceeded satisfactorily and the reservoir was opened in 1933. One highly attractive feature is the concrete bell-mouth overflow system. Further up the valley is the Ben Crom Reservoir, designed by Binnie, Deacon and Gourley and built from 1953 to 1957 by Charles Brand & Co. It is a mass-concrete gravity structure and at 125ft it has the greatest free height above the riverbed of any dam in Ulster. (HEWs 1889, 2119)

The Silent Valley Reservoir
LIAM LYNCH

U25. White and Seafin Bridges. The Upper Bann River is crossed by two cast-iron bridges with brick jack-arching. This form of bridge was never as common in Ireland as it was in Britain. It is therefore unusual to find two extant bridge decks of this type situated relatively close together.

White Bridge, (J 048499), was built by Courtney, Stephens & Co. of Dublin, most likely in the 1860s. It has three longitudinal cast-iron I-beams 24in by 12in supporting jack-arches. In 1985, the undersides received an application of gunite, and the original fill over the brick jack-arches was replaced with lightweight concrete. The bridge has four spans, each of 32ft, and is supported on cylindrical columns with only a light cross-bracing near the underside of the deck. The road is 14ft wide between modern balustrades, which replaced the original ornate cast iron railings following a road traffic accident. (HEW 2141)

Seafin Bridge (J 216380) was erected in 1878 by R. Lucus & Sons, Soho Foundry, Newry. This bridge has three spans of 20ft. On this structure, the brick jack-arching, balustrades and deck are as built; the supports, however, are concrete block piers. (HEW 2134)

White Bridge

U26. Killough Windmill (J 536360) was built by Lord Bangor in 1823 and was recorded as having two pairs of 'Donegal' stones (quartzite conglomerate), one for oats and one for wheat, together with a mill cottage and a kiln.

In 1852, Lord Bangor installed flax-scutching machinery, making it unique in having flax-scutching and grain-milling operations in the one windmill. The scutching operations ceased around 1857 and corn milling in 1859/60. The mill is essentially a composite design of a 'podium' tower mill (of which there are eight examples in England) and a 'vaulted' tower mill (eighteen of which are found in Scotland) and is believed to be unique in Britain and Ireland, having a storage area surrounding its base, constructed in masonry. The 1903 OS map indicates that the mill was in ruins. Only the stonework and the covered roof to the undercroft remain, all remnants of the timber floors, machinery, millstones, roof and sails no longer being on site.

This mill is an indicator of the buoyancy of the agricultural industry in the early nineteenth century in the area and is an important monument to the innovative nature of the Ward family (Lord Bangor) and their efforts to promote agriculture in the region. (HEW 2649)

Killough Windmill

U27. Nendrum Tide Mill (J 525637). From early medieval times until the industrial revolution, tide mills existed all around the western coasts of Europe and further afield. They functioned by trapping water at high tide and then, when the tide had fallen, releasing the water to drive a horizontal waterwheel, which in turn was connected by a shaft to a millstone which ground corn.

Strangford Lough is a tidal lough 15 miles long and 4 miles wide, lying southeast of Belfast. Almost 1,400 years ago, the monks of Nendrum Monastery on Mahee Island in Strangford Lough were extracting energy from the tidal range to drive their flourmill. Archaeological excavations took place at Nendrum in 1844 and 1922–24, so much is known

Remains of dam across the inlet at the site of the Nendrum tide mill

about the monastery and the way of life there in medieval times, but there was no evidence of a tide mill.

In 1995 an Order for Northern Ireland permitted archaeological exploration from high-water mark to a distance of 12 miles out to sea. A two-week excavation was started on the foreshore at Nendrum in April 1999 and nothing of note was discovered until the last afternoon of the dig when two small items were found which changed the whole understanding of the site. They indicated that Nendrum had had a tide mill. Exploration continued until December 1999 and again in 2000 and 2001. With the advent of accurate tree ring dating in 1970, particularly of oak trees, it was possible to date the construction of a second mill at Nendrum to AD 789 and the first to AD 619, which makes it the earliest known tide mill in the world.

From the archaeological evidence collected it was determined that water trapped at high tide was fed through a 13ft-long rectangular duct formed from a block of sandstone with a close fitting lid. The solid outlet end had a 5in circular hole through which the water drove a 3ft-diameter wheel with twenty-four spoon-shaped paddles each 1ft long. The wheel had a vertical shaft 8ft 3in long, which was attached to an upper millstone made from Newry granite. The remains of the dam across the inlet can still be seen. (HEW 2354)

U28. Hunt's Park, Donaghadee and Downpatrick Water Towers. Both water towers, now disused, are structurally exceptionally well preserved. They were constructed by the Expanded Metal Company using expanded metal reinforcement in 1912. These towers are important as they are still intact and an early example of the use of reinforced concrete. The condition of the concrete is good and shows no sign of rust or spalling, nor of any honeycombing and patching up, which is a tremendous tribute to the contractors of the time. Access to the towers is through a single metal door set in one of the semicircular arch recesses, which are formed in the battered base. The tanks have a simple parapet and conical roof. The contractor for the Hunt's Park tower was J. & R. Thompson of Belfast, whilst the tower at Downpatrick was built by R. D. Pollock

The larger of the two towers in at Donaghadee, (J [illegible]), which has an overall height of 70ft. The tower is 18ft in diameter supporting a 24ft (internal) diameter tank some 12ft deep, with 10in-thick walls. The tower was commissioned by Donaghadee Urban District Council providing the town with its first mains water supply and remained in use until the 1940s when it was replaced by a new reservoir at Orlock. The tower is secured with a locked steel door and is set in a well-kept park. (HEW 2113)

The second tower is situated in the grounds of the asylum at Downpatrick (J 500442) and has not been in use for many years. This structure is 55ft tall, with a 16ft-diameter tank some 11ft deep supported on a 13ft column. In this tower the reinforcement can be seen in a recently cut opening, and was confirmed as ¼in bars in a 2in-square mesh,

Water tower at Hunt's Park, Donaghadee

positioned, not at the face, but in the centre of the concrete wall. Whether this positioning of the mesh was deliberate or a fortuitous error will never be known, but it has probably been the main reason for the undamaged longevity of both structures. (HEW 2114)

U29. The Lagan Navigation (Sprucefield–Lough Neagh) (J 079627 to J 260626). The original intention of the Lagan Navigation was for a canal from Belfast to Lough Neagh, and work began from Sprucefield in December 1782, supervised by Richard Owen, and largely financed by the Marquis of Donegall. The main structure on this stretch of canal was the 300ft-long aqueduct over the River Lagan at Spencer's Bridge, on which work commenced first. It took three years to complete and cost £3,000. The new cut was formally opened on New Year's Day, 1794. It started at Sprucefield with a flight of two double locks (the Union Locks) giving a rise of 26ft to the summit, which ran for 8 miles to Moira. There were then ten locks (each 70ft by 16ft) taking the canal down to Lough Neagh. Although the most successful of the canals in Ulster, the Lagan Navigation eventually lost out to other forms of transport. It was abandoned above Lisburn in 1954 and totally in 1958. The canal bed between Sprucefield and Moira was used for the line of Ireland's first motorway (the M1) and the aqueduct was demolished. However, the line of the cut beyond the aqueduct remains and a number of fine bridges still exist, including a superb skew brick arch just beyond Moira Station. (HEW 2117/2)

Lock 20 and Sheerin's Bridge – small arch bridge for the horse

1730. Crawfordsburn Viaduct (J 466817) is on the Belfast–Bangor railway line spanning across a 70ft deep gorge in the picturesque Crawfordsburn Regional Park. The five-span structure was constructed in locally quarried Carboniferous limestone with brick vaulting in the 46ft-span arches.

The design was by Charles Lanyon who also designed the main Bangor station and several smaller stations on the line. The foundation stone was laid in 1863. The portion between Holywood and Bangor was constructed in single Irish gauge (5ft 3in) track, when most of the bridges were constructed in double track width. The second track was eventually laid in 1902, when the stone parapets were replaced by a steel deck which was changed to a precast concrete deck in 1978–79. The total length of the viaduct is around 330ft with a gradient of 1 in 107 rising towards Bangor. The old jointed track was replaced in 2006–07 by continuous welded track. (HEW 1974).

Crawfordsburn Viaduct

BELFAST CITY & DISTRICT

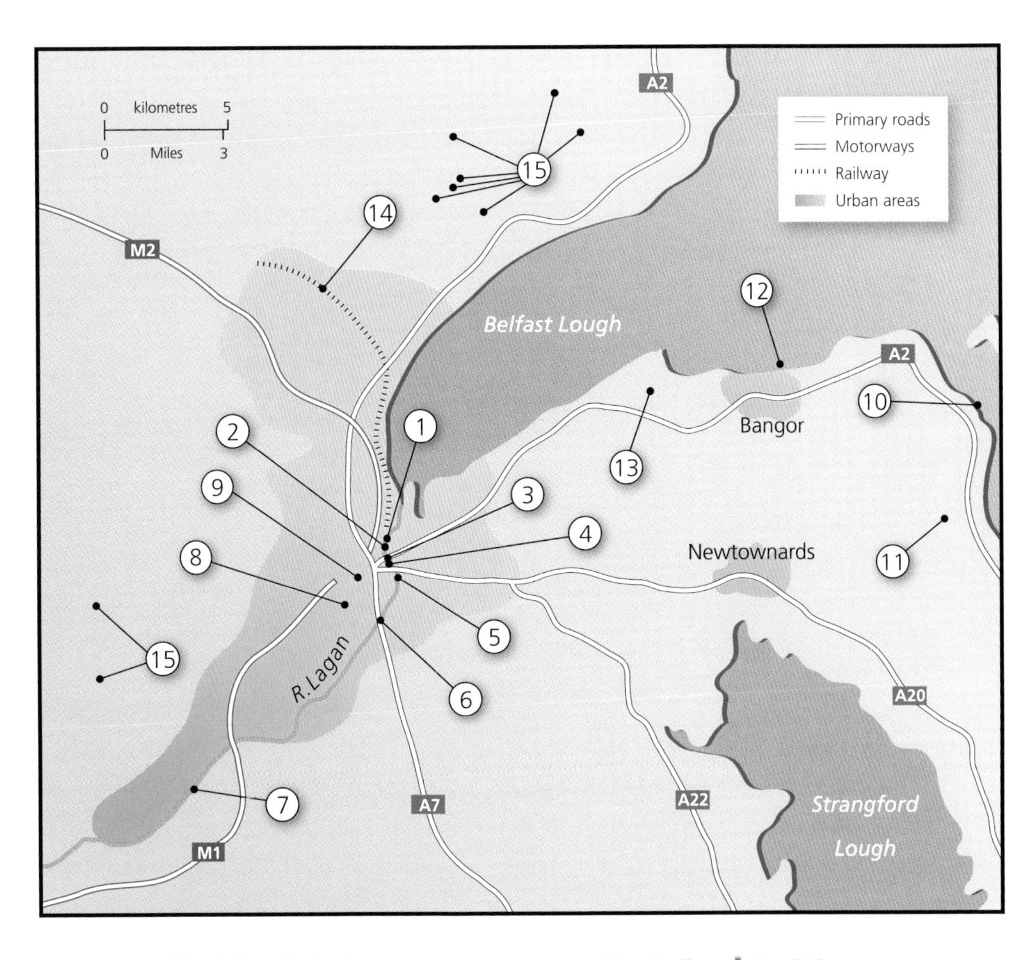

B1 Belfast's Dry Docks
B2 The Cross Harbour Bridges, Belfast
B3 The Lagan Weir, Belfast
B4 Queen's Bridge, Belfast
B5 Albert Bridge, Belfast
B6 King's Bridge, Belfast
B7 Lagan Navigation (Belfast [illegible]
B8 Tate's Avenue Bridge, Belfast
B9 The Boyne Bridge, Belfast
B10 Donaghadee Harbour & Lighthouse
B11 Ballycopeland Windmill
B12 Bangor Breakwater
B13 Helen's Bay Station
B14 Greenisland Loop Line
B15 Belfast's Reservoirs

B1. Belfast's Dry Docks. The original port of entry to County Antrim was at Carrickfergus, and access to Belfast remained restricted until William Dargan cut the first access channel in 1839–41. The shipbuilding industry began on a small scale in 1791, and was able to progress much more rapidly when proper access was provided and following the construction of a series of dry docks (see Table). The Clarendon docks were on the County Antrim bank of the River Lagan, but in 1853 shipbuilding commenced on Queen's Island (formed on the County Down shore from the spoil dredged by Dargan). This became the nucleus of the extensive shipyards of Harland and Wolff.

Table 1: Early Belfast Dry Docks

DOCK	HEW	GRID REF NUMBER	YEAR OPENED	LENGTH (ft)	WIDTH MIN/MAX (ft)	SILL DEPTH BELOW HD	DESIGNER	CONTRACTOR
Clarendon 1	1598	J 344751	1800	235	28	at datum	Wm Ritchie	Wm Ritchie
Clarendon 2	1599	J 344751	1826	285	34	at datum	Thos Burnett	Mullins & MacMahon
Hamilton	1807	J 351752	1867	450	60/64	5ft 7in	Lizars	Thos Monk
Alexandra	1816	J 355759	1889	788	50/80	15ft	Salmond	McCrea & McFarland
Thompson	1819	J 357762	1911	850	96	24ft 6ins	Harbour Comms (in house)	Scott & Middleton

Of the works listed in the Table, both of the Clarendon Docks are now preserved in a dry state, and the Alexandra Dock is host to HMS *Caroline*. On the Hamilton Dock, now host to SS *Nomadic*, a young L.L. Macassey was the contractor's site agent. It will be seen that each dock constructed was progressively larger. When it was built, the Thompson Dock was the largest in the world and was used by the *Olympic* and the *Titanic*. (As the integrity of the lock gate was in doubt, a concrete wall has recently been built outside the gate of the Thompson Dock, which enables members of the public to gain access safely to the floor of the dock). By 1962, even the Thompson Dock was too small for the large tankers and bulk-ore carriers then being built and work commenced on an even larger dry dock.

The Belfast Dry Dock (HEW 1853) (J 362769) was designed by Rendel Palmer and Tritton and built by Charles Brand & Sons from 1965 to 1968. Whilst all of the earlier dry docks are in stepped masonry, the new dock consists of a simple reinforced concrete floor, anchored against uplift, and sheet pile walls. It is some 1150ft long by 160ft wide. When constructed, it was one of the five largest in the world. The final major construction work undertaken was the Building Dock (HEW 2109) (J 356752). The walls are L-shaped reinforced concrete cantilevers carried on steel piles driven to bedrock. It was completed in 1970 by George Wimpey & Co. to a design prepared by Babtie, Shaw & Morton. The Building Dock marked a change in shipbuilding practice. Large ship sections could now be built under cover using automated welding and then taken to the dock for assembly.

Thompson Dry Dock

Two large Krupp gantry cranes were installed. Dominating the skyline, these are affectionately known as *Samson* and *Goliath*. They have a span of 153ft and a headroom of 227ft and 266ft respectively. Each has a safe working load of 840tons. Sadly, shipbuilding has now ceased in Belfast. However, the facilities are still in use for ship repairs and the fabrication of wind turbines.

B2 The Cross Harbour Bridges, Belfast (J 344745) represented the largest single transportation project undertaken in Ulster. The work was awarded as a design-and-build contract in 1991 to a joint venture formed by John Graham (Dromore) Ltd and Farrans Construction Ltd, Belfast. The main design was undertaken by Acer Consultants Ltd and was approved by the Royal Fine Art Commission. There are actually two bridge structures

Cross-Harbour Bridges, Belfast

side by side: the road viaduct, 800m long, connects the Sydenham bypass to the Westlink (and hence to the M1 and M2 motorways); the rail viaduct joins the previously isolated railway from Larne to the rest of the system at Central Station. The main road structure, the Lagan Bridge, crosses the river in three spans, an 83m central span flanked by 55m side spans. The rail structure, the Dargan Bridge, is 1,490m long, making it the longest bridge in Ireland. Work was completed in August 1994 at a cost of £32 million. The viaducts are supported on 105 land piers, and two sets of river piers, all piled. The decks consist of 1,058 matched precast segments, weighing between 50 and 90 tonnes each, built off the piers as balanced cantilevers. This was the first time that match casting (each unit being cast against its neighbour in the off-site casting yard) was used in Ireland. The units were post-tensioned after placement. (HEW 2115)

B3. The Lagan Weir, Belfast (J 344744). The organic nature of the slob lands in Belfast Lough caused environmental nuisance for many years. In the 1920s, W.J. Campbell & Sons was employed to construct a weir and barge lock across the River Lagan upstream of its confluence with the River Blackstaff. Known as the McConnell Weir, it incorporated a 40ft-long sluice gate. The works were completed in 1937 at a cost of £358,000. Siltation upstream of McConnell Weir, and continuing problems caused by the slob downstream, led to a reassessment of the situation in the 1980s. Following studies undertaken in 1988, using a large model constructed by the Queen's University of Belfast, a new weir was designed by Ferguson and McIlveen and built by Charles Brand & Sons Ltd. It is located between the Queen Elizabeth Bridge and the Cross-Harbour Bridges. The new Lagan Weir has two functions. First, it holds sufficient water in the river upstream to prevent unsightly mud banks from being uncovered at low tide. Secondly, by adjustment of the curved gates, it can hold back a flood tide downstream. Because of the nature of the slob land, extensive foundation work was required for the four river piers. The five 20-tonne gates were constructed in the nearby yards of Harland and Wolff. The work was completed in 1994 at a cost of £14 million. Once the Lagan Weir became operational, the gate at the McConnell Weir and the sill were removed, the rest of the structure being retained as a feature. (HEW 2116)

The Lagan Weir, Belfast

B4. Queen's Bridge, Belfast (J 344743). The Long Bridge, some 760ft in length with 1,800ft of approach causeway, built between 1682 and 1688 on the site of a ford over the River Lagan, was, by the end of the eighteenth century, in a dangerous state. Accordingly, the grand juries of Antrim and Down obtained a loan of £28,000 from the Board of Works for a replacement. By infilling on the County Down bank, the spans of the new bridge, built by Francis Ritchie, were reduced from twenty-one to five flat segmental masonry arches, each of about 50ft span. It was opened in January 1843. Although built to a width of 40ft, traffic growth resulted in a need for more lanes. In 1886, the city surveyor, J.C. Bretland, designed 7ft-wide cantilevered footways on either side. These are supported by metal brackets carried at the piers by clusters of columns with large capitals. In 1966 a new parallel bridge, with one major span and half-span side spans, was built in steel. When this, the Queen Elizabeth II Bridge, was opened, the Queen's Bridge (named after Queen Victoria) became one-way. (HEW 2100)

B5. Albert Bridge, Belfast (J 349738). At this site in 1831 a bridge was opened which carried a toll of one halfpenny, payable by everyone crossing. In 1860, it was purchased by the local authorities for £4,500 and the name changed to the Albert Bridge. The increase in traffic caused structural difficulty and in 1886 it was noticed that part of the bridge was sinking; some 70ft of the centre collapsed suddenly on 15 September, fortunately at night. A replacement bridge, designed by J.C. Bretland, was opened on 6 September 1890. This has three flat segmental arches, each with a clear span of 85ft and each made up of eleven cast-iron ribs. Each arch rib is made up of five segments with intermediate radial stiffeners. Although of a relatively late date, this bridge across the River Lagan is one of the finest in Ireland built in cast-iron. The abutments are in brick with granite ashlar facings. The contractor was James Henry & Sons of Belfast and the castings were by Handyside & Co. of Leeds. The contract price was £36,500. In 1985 a barge collided with the bridge, causing damage to the outer arch rib on the upstream face. The Roads Service DOE (NI) awarded a repair and restoration contract, with Dr I.G. Doran & Partners acting as consulting engineers. The elements of the bridge are now picked out with paints of contrasting colours. (HEW 1020)

B6. King's Bridge, Belfast (J 340718). From 1869, the River Lagan in Belfast was crossed at only four points (one rail bridge and three road bridges). By the early 1900s this was proving to be inadequate and in 1909 Belfast City Council advertised for designs for another road crossing. The competition was won by the Trussed Steel Concrete Co., which offered a four-span concrete bridge reinforced on the Kahn principle, a system that originated in the USA for large factory buildings, but by 1909 was being promoted for bridge work. The Kahn reinforcing bar was diamond in section and had horizontal extensions which could be bent up or down to provide bond or resist shear. The new bridge, opened in 1912 and

Above: Queen's Bridge, Belfast

Below: Albert Bridge, Belfast

named after the new king, is the earliest multi-span Kahn bridge identified in either Britain or Ireland. The distance between the abutments is 195ft and the width is 30ft. Minimum headroom is 9ft 6in. There are three main beams, 51in by 13in section for the two 50ft central spans, reducing to 42in by 12in for the outer 40ft spans. It was constructed by W.J. Campbell & Sons. (HEW 1968)

B7. Lagan Navigation (Belfast–Lisburn) (J 341712 to J 261629) Work on a scheme to make the River Lagan navigable from Belfast to Lisburn commenced in 1756 supported by grants from the Irish parliament and directed by Thomas Omer. The navigation was opened to Lisburn in 1763 and extended two years later to Sprucefield. The course of the river was followed throughout. Twelve locks were needed, giving a total rise of 81ft 9in. Lock 3 has recently been refurbished and the lock-keeper's cottage is open to visitors.

Between locks 3 and 4 is Shaw's Bridge (J 325691). This site was for many years the main crossing point on the River Lagan for the road from north County Down to Dublin. Petty's map of 1657 shows a timber bridge at the site, which was built during the Cromwellian wars in 1655 by a Captain Shawe to pass his ordnance. In 1698 Thomas Burgh rebuilt it in stone, but this bridge was swept away in 1709. The Grand Jury rebuilt it, but it too was swept away. Rebuilt again, and repaired in 1778 and 1856, there are currently five near semicircular arches, one of 23ft 6in and four of approximately 30ft. The pier thicknesses at 7ft to 8ft have triangular cutwaters up to the level of the arch springing. The width between parapets is 16ft 3in. The bridge is now restricted to pedestrian traffic and has been bypassed by a single-span concrete arch bridge, built in 1974 a few metres to the east. (HEW 1969)

King's Bridge, Belfast

Shaw's Bridge, Belfast

Between locks 6 and 7, at Drum (J 307672), is a fine example of a causeway bridge as used by early road builders to cross the flood plain of a river. (A raised causeway, with stone paved sides, was constructed and the river was then crossed by a conventional, usually multi-span, bridge.) The causeway at Drum is approximately 510ft long. The three main arches are semicircular and have spans of 19ft, 22ft and 19ft respectively. The width between the parapets is 31ft. The causeway at Drum was later pierced by the Lagan Navigation. The edges of the road on the Belfast side of the bridge are defined by fine examples of stone walls and on one parapet is the pulley used for the towrope when the towpath changed from one bank to the other. (HEW 2118)

At Ballyskeagh (J 289668) a tall, elegant two-arch bridge, known as the High Bridge, was built in sandstone. Thomas Omer designed there a standard lock-keeper's house, and while a number remain in various states of disrepair two have been refurbished, at Ballyskeagh and at Drum. At the Island Civic Centre in Lisburn, lock 12 has been rebuilt. (HEW 2117/1)

Lagan Navigation

R8. Tate's Avenue Bridge, Belfast (J 324726). The bridge at Tate's Avenue is a remarkable mixture of styles and designs. It was built for the Belfast City Council in 1926 to replace a level crossing over the Great Northern Railway (Ireland). The bridge is, in essence, three structures comprising a ramp, a series of steel spans and a further ramp. From the east (Lisburn Rd) end there is a 408ft-long ramp, a span of 39ft (over a road), a span of 60ft (sheeted in for storage), the 78ft main span over the railway, a span of 54ft (sheeted in for storage), a span of 39ft (over a road), a span of 42ft (sheeted in), and, finally, a second ramp at the west end. The spans comprise longitudinal riveted steel plate girders of varying depth with transverse girders with brick arching between. The footways are carried on cantilevered girders, again with brick arching between. Notable are the massive scrolls, cantilevered from brick piers, which support the outer girders on the road spans. The overall width is 44ft. The main contractor for the bridge was H. & J. Martin. The heavy steel work was undertaken by the Motherwell Bridge and Engineering Co. The iron handrail supports, which carry lattice-work handrails, were cast by the Millfield Foundry, Belfast. In the early 1990s Dr I.G. Doran & Partners carried out an extensive survey of the structure. Overall cleaning and repainting was carried out in 1992. To protect the metal parapets from vehicle impact, crash barriers were added at each side of the roadway. (HEW 2143)

Tate's Avenue Bridge, Belfast

B9. The Boyne Bridge, Belfast (J 334737). The old road from Carrickfergus to Lisburn crossed the Blackstaff River at its tidal limit by the Salt Water Bridge, which was built in 1642 and extensively repaired in 1717. The terminus for the Ulster Railway was constructed adjacent to the bridge. However, by 1870, the level crossing on the approach was causing considerable delays and the railway company built a three-span road bridge in wrought iron abutting the Salt Water Bridge. In 1934 R.B. Donald, Belfast City Surveyor, designed a wider replacement structure. The resulting bridge, known as the Boyne Bridge (reputedly because, in 1689, the Williamite Army used the Salt Water Bridge on its way to the Battle of the Boyne) is an interesting composite structure. On the southern approach, parts of the stonework of the old bridge remain, now covered by a layer of reinforced gunite (sprayed concrete). The bridge is 60ft wide by 138ft long and is made up of eight plate girders supporting the carriageway and two parapet plate girders. The total weight of steel is 300 tons. The foundation consists partly of precast and partly of cast *in-situ* piles and the north abutment was designed as a cellular lightweight concrete construction. The deck is formed by an 8in-thick concrete slab. The contractor was H. & J. Martin, and the work was completed in December 1936. (HEW 2092)

The Boyne Bridge, Belfast

B10. Donaghadee Harbour and Lighthouse (J 591001) lie at the northeast corner of County Down. The shortest crossing on the Irish Sea is between Portpatrick and Donaghadee. In the early seventeenth century Viscount Montgomery owned lands in both areas and by 1606 he had started to build harbours on each side of the Irish Sea

The first harbour at Donaghadee was probably a timber landing jetty built around 1610 and improved in stages up to 1662. By 1774 a second larger harbour had been designed

by John Smeaton, the leading civil engineer of the time, and was constructed at a cost of £10,000. This harbour extended the earlier construction eastwards with a curved breakwater that turned north.

At the beginning of the nineteenth century, larger vessels were having difficulty in berthing at low tides. John Rennie senior proposed an enlarged harbour, extending the 1774 pier further out into the sea in an easterly direction, and introducing an isolated breakwater to the north and a new lighthouse on the south wall. Construction started in 1821 and was supervised by Rennie's son, Sir John Rennie. The outer faces of the piers are constructed from local greywacke stone at a slope of 1 in 5, and the inside faces of the piers and the pier heads from limestone quarried in Anglesey. (At low tide, the 1774 breakwater base can be seen covered in seaweed on an otherwise sandy seabed.) In the 1970s, the 1610 harbour was filled in to form a children's play area.

Ferry traffic to Scotland eventually could no longer be catered for, and moved in 1867 to the more sheltered harbours of Stranraer in Loch Ryan, and Larne in County Antrim. The lighthouse, commissioned in 1836, was electrified in 1934, making it the first lighthouse in Ireland to use electricity from mains supply.

Rennie's harbour opened in 1825 and is little changed. This is a well-preserved example of a Georgian harbour, constructed to handle sailing vessels, complete with warping rings recessed into the quay face. The harbour is used by local fishermen with their small boats, by leisure craft, and by holiday pleasure-trip boats. It supports a Royal National Lifeboat Institution station, which celebrated its hundredth birthday in July 2010. (HEW 2110)

Donaghadee Harbour and Lighthouse

B11. Ballycopeland Windmill (J 580761) lies in the northeast corner of County Down, 2 miles south of Donaghadee, and is the only complete windmill in Ulster. The tower, believed to date from about 1784 is built of local Silurian gritstone, some 33ft high and tapering from 22ft at the base to just under 14ft at the top. In 1937 it passed intact into state ownership and is maintained by the Northern Ireland Environment Agency. The windmill has a boat-shaped revolving roof supporting the wind shaft, which carries the four sail arms, and an eight-bladed fantail designed to turn the sails into the wind. The motion is carried to the vertical timber shaft by a bevel gear on the wind shaft, and then through a series of timber toothed gears to the various millstones and bag hoists. A special point of interest is the mechanism to reef the sail arms depending on the wind strength – 'Hooper's Patent Reefing Gear' – that may be unique in Britain and Ireland. The mill originally produced wheaten, oaten and maize meals using three sets of stones.

Also on the site are the kiln and the former miller's house representing the last of a large number of similar sites on the east coast of County Down, once known as the grain basket of Ulster, all of which have either been demolished, allowed to decay, or have been converted into dwellings. (HEW 2086)

Ballycopeland Windmill

B12. Bangor Breakwater (J 503822). Bangor lies on the southern shore of Belfast Lough and like most small harbours in Ulster, developed over the years, with the first pier being built in 1757. To counteract wave action caused by northerly winds, the North Pier was constructed in 1895, but the harbour still needed protection from the west.

In 1977, North Down Borough Council commissioned a study of a seafront development scheme, which was to include a marina. The consultants, Kirk McClure Morton in conjunction with the Queen's University of Belfast, carried out extensive model studies, as a result of which it was decided to partially rebuild and extend the existing North Pier by 150m and to construct a new pier to the west, some 250m long, to form an enclosed harbour.

Conventional rubble breakwaters were considered, but there was a problem at the time with the availability of 8–12 tonne boulders required for the outside face of the construction. Instead, precast-concrete 1.2m cube units were developed in conjunction with the Hydraulics Research Station at Wallingford, Oxon. The hollow, energy-absorbing concrete units form the outer face of the breakwater, rising from a cast *in-situ* concrete toe beam that is visible at low tides.

The work commenced on the project in 1979 and the marina was finally completed in 1994. The appearance of the SHED (Shephard Hill Energy Dissipator) units produce a pleasing and rather unusual finish to the breakwaters. (HEW 2095)

Bangor Breakwater

Helen's Bay Station

B13. Helen's Bay Station (J 458821) is on the main Belfast–Bangor railway line, the twin-track 5ft 3in gauge line serving a small town on the southern shore of Belfast Lough adjacent to Crawfordsburn regional country park. The Marquis of Dufferin and Ava has his estate at Clandeboye, County Down. From the house a private road for carriages was constructed to Grey's Point, some 3 miles away. This crossed over or under the public roads encountered along the way, the underbridges being decorated with carved ornamentation, the overbridge being of simpler construction. With the construction of the railway to Bangor, which also had to cross the private road, a very ornamental bridge was built adjacent to the station, which itself is a miniature example of the Scottish Baronial style. Two projections from each face of the bridge pass through the platform walls. These were designed by architect Benjamin Ferrey in a style considered 'suitable' and the bridge was opened in 1864. The building, platforms and bridge have been listed since 1975 as Grade 'A' structures by the Northern Ireland Environment Agency. (HEW 2112)

B14. Greenisland Loop Line (J 353833 to J 315854) contains the greatest concentration in Ulster of concrete structures built specifically for the railways. The original railway north of Belfast served the town of Carrickfergus. From near Greenisland Station, the railway to Ballymena branched, going west and then north. Trains from Belfast to the west had to stop at Greenisland, where they either reversed, or the engine was run round the train, causing considerable delays and expenditure. Between 1931 and 1933 the Northern

Counties Committee of the LMS Railway Co. constructed a 2½-mile cut-off line. This was designed to allow trains to the west to proceed unimpeded by services to Larne. The whole scheme was undertaken by direct labour as an unemployment relief scheme. The works include four under- and four overbridges, all different, and two large viaducts, all constructed in concrete. The largest structure, which was claimed to be the largest reinforced concrete railway viaduct in either Britain or Ireland when built, is the main line viaduct. This has three main arches (83ft clear spans) and nine approach arches (35ft spans). At the Belfast end is an additional rectangular opening, which passes the down Larne line underneath as a burrowing junction, being the only one in Ireland. The down line then crosses the down-shore viaduct (one main arch and six approach arches). These two structures contained 17,000 cu.yds of concrete. To the south, the railway tracks cross a road by three parallel square profile bridges, each carrying a different year date. Continuing north are four bridges of differing design – Jordanstown Road Bridge, Monkstown Bridge, Auld's Bridge and Mossley Bridge. The different designs may be regarded as something of an experiment and these works therefore represent an important stage in the development of the use of reinforced concrete for railway work. (HEW 2111)

Valentine's Glen Viaducts, Greenisland

B15. Belfast's Reservoirs The demand for water, as a result of the rapid growth of Belfast from the mid-nineteenth century, soon outstripped the inadequate water supply then available. To improve the situation, a series of surface impounding reservoirs was constructed in the County Antrim hills above Carrickfergus and Lisburn. These were built in accordance with proposals that were first formulated by J.F. Bateman in 1855 and continued by L.L. Macassey from 1874 until 1891, when the Mourne scheme was proposed by Macassey.

Construction was generally of earthen embankments with clay puddle cores and all but two are classified as 'large'. All the dams are straight, with the exception of Dorisland and Stoneyford, which are U-shaped in plan. Despite the age of these reservoirs, all still play their part in the supply of water to Belfast. (HEW 2137)

NAME OF DAM	GRID REFERENCE	DATE OF ACT	DATE OPENED	LENGTH (ft)	HEIGHT (ft)	RESERVOIR CAPACITY (million gallons)	CONTRACTOR
Woodburn North	J 371913	1865	1868	2060	41.5	82	J.D.Cooper
Woodburn Middle S.	J 371891	1865	1868	1780	84	469	J.D.Cooper
Woodburn Lower S.	J 378892	1865	1880	1055	61	10	Peter Quin
Woodburn Upper S.	J 364888	1874	1876	1615	71.5	367	J.Gilmer + DL
Dorisland	J 384881	1874	1878	1070	41.5	66	David Smith
Lough Mourne	J 413928	1879	1879	630	14.5	576	No embankment built
Copeland	J 428915	1879	1879	2030	61.5	133	WJ Doherty + McCrea & McFarland
Stoneyford	J 218697	1884	1889	3150	38.5	810	Anwell & Partners + DL
Lagmore	J 254666	1884	1889	800	30	21	Fitzpatrick Bros
Leathemstown	J 215726	1889	1892	1460	62.5	99	DL (Direct Labour)

Table 2: Belfast's reservoirs

Dorisland Reservoir

GLOSSARY

abutment – the structure (frequently masonry) at each end of an arch, or series of arches, which provides the resistance to the thrust of the arches; or, more generally, the supports at the ends of a bridge.

aqueduct – a bridge-like structure used to carry a watercourse, frequently a canal, over a river, road or railway.

arch – a curved structural form used to support the deck of a bridge. Can be built in stone, concrete or steel.

arch types:

elliptical – similar to segmental but with the centre portion flattened;

pointed – two half-segments with radii displaced to opposite sides of arch, meeting at the centre at a sharp point, can be sharp or flat;

relieving – an arch built into a structure to reduce the load on a wall below;

semi-circular – formed of one half of a circle, rise equal to half the span;

segmental – formed of less than half a circle, rise less than half the span.

balanced cantilever construction – the bridge is built in increments on each side of a pier. Construction can be either by casting the increments *in situ* or by casting the sections off-site and transporting and lifting them into place. The method is particularly useful where access under the bridge is difficult. Also used for steel bridges.

bascule – a single or double leaf bridge usually over a waterway which can be raised by pivoting about a horizontal axis to give headroom for vessels to pass through.

bending moment – the product of a force and a distance measured at right angles to the force.

berm – a narrow ledge or shelf

blacktop: a collective term used for all forms of modern road surfacing

asphalt – a high strength road surfacing material of stones with a binder of bitumen or natural asphalt;

bitumen – a by-product of distillation of crude petroleum oil;

tarmacadam or **tarmac** – a road surfacing material of stones with a binder of tar.

blinding – a layer of fine sand or gravel spread over a road surface to fill cracks.

bowstring girder – a girder with a straight bottom chord from which springs an arched top chord, the two connected by steel wire cables or rigid stiffening members.

box girder – a girder of rectangular or trapezoidal cross section whose webs and flanges are relatively thin.

buckled plate – a cast- or wrought-iron square or rectangular plate having a convex form to resist downward loads.

cable-stayed bridge – a bridge whose deck is directly supported by a fan of cables from the towers.

caisson – A watertight structure within which construction work is carried on under water, sometimes using compressed air.

canal: a man-made inland waterway principally for barges

inland navigation – a combination of canal and river;

lateral canal – a short length of canal with usually one lock which bypasses rapids or shallow water in a river;

summit level canal – a canal which has lower levels in both directions from the summit length;

pound lock – an enclosed chamber with gates at both ends for moving boats from a higher to a lower level and vice versa;

flash lock – a single gate in a weir that can be opened to allow boats to pass through.

cantilever – a beam not supported at its outer end.

cast iron – an iron-carbon alloy with impurities which preclude it being rolled or forged; it has to be poured in molten form into moulds of the required shape and size.

catenary – a curve formed by a wire, rope, or chain hanging freely from two points that are on the same horizontal level.

cofferdam (or coffer) – an enclosure within a water environment constructed to allow water to be pumped out to create a dry working environment.

centring – a temporary structure to support an arch during building.

continuous beam – a beam of several spans in the same straight line joined together in such a way that a known load on one span will produce an effect on the others.

cut and cover – a tunnel or part of a tunnel constructed by excavation a cutting to build it and then burying it; often used at the ends of a tunnel to reduce the amount of actual tunnelling required.

cutwater – a V-shaped upstream face of a bridge pier designed to part the flow of water and prevent debris accumulating around the pier; sometimes provided also on the downstream face, for aesthetic or structural reasons.

dam: a structure built to contain the water in a reservoir

embankment dam – resists the water pressure by its own weight. Can be of compacted earth or rockfill with a vertical impermeable central puddle clay core to prevent seepage through the dam. This form is particularly suitable for wide flat valleys. Rockfill types can have an impermeable upstream face rather than a core. A cut-off trench filled with puddle clay or concrete is also required under the dam to prevent seepage;

gravity dam – resists the water pressure by its own weight. Constructed of masonry or concrete, but because of the greater mass of the material the downstream and upstream faces can be quite steep. This form is suitable for narrower valleys with sound rock for a foundation;

buttress dam – a development of the gravity dam with the downstream face supported by buttresses, thereby reducing the amount of material necessary to resist the water pressure;

arch dam – of masonry or concrete, used where the rock formations at the sides of a valley are able to support the thrust from the arch form. Sound rock is also needed for the foundations. This form of dam is capable of being built to greater heights than an embankment or gravity dam.

extrados – the convex surface of an arch

formwork – (or shuttering), a temporary structure to contain wet concrete and form its finished shape until it has set and is self-supporting.

gabions – stone-filled wire baskets used for batter protection.

girder – a beam formed by connecting a top and bottom flange with a solid vertical web.

grout – a mixture of cement and water.

grouting – the filling of the voids in structural elements with grout or a fluid mortar.

historical periods:

early medieval – AD 410 to 1066;

medieval – 1066 to 1536;

post-medieval – 1537 to 1900;

modern – 1900 to present day.

immersed tube tunnel – prefabricated units in steel or concrete laid in an excavated trench across a river bed, joined together and covered over to form a continuous tunnel.

interceptor sewer – Large sewers that, in a combined system, control the flow of sewage to a treatment plant.

intrados – the inner curve of an arch, also known as the soffit.

invert – the lowest level at which fluid flows in a pipe.

keystone – the central voussoir, often decorated.

littoral drift – the movement of sand or shingle along a coastline by the action of tides striking the coast at an angle and scouring the beach material.

loads: the forces applied to a structure

dead load – the self weight of a structure and any permanent attached loads

live load – variable loads, such as traffic crossing a bridge, and loads from wind, snow, water etc.

mass concrete – concrete without added steel reinforcement.

mild steel – steel containing a small percentage of carbon, strong and tough, but easily tempered.

mortar – a mixture of cement, sand and water used in brickwork and to fill the voids in masonry. When it has hardened the mortar distributes the pressures exerted by loads.

OD – Ordnance Datum, the mean sea level at Malin Head, County Donegal, from which Ordnance Survey Ireland and Ordnance Survey Northern Ireland measure heights in Ireland.

pavement – the main traffic-carrying structure of a road

flexible – a pavement constructed of asphalt or tarmacadam, or stone or concrete blocks;

rigid – a pavement of mass or reinforced concrete.

pattress plate – metal spreader plate connected by a tie rod inserted through the spandrels of a masonry bridge to prevent them from moving outwards

penstock – a valve controlling the flow of water in a large diameter pipe.

pilaster – a rectangular column, partly built into a wall and partly projecting from it.

portal frame – a frame of two vertical members connected at the top by a horizontal member or two inclined members with rigid connections between them.

prestressed concrete – concrete into which compressive stresses are locked so that the concrete can resist tensile stresses due to live loading

pre-tensioned concrete – concrete steel bars or wires which are tensioned before the concrete is cast;

post-tensioned concrete – concrete in which steel bars or cables are contained in ducts cast in the concrete and which are tensioned after the concrete has hardened by jacking the bars or cables against anchorages cast into each end of the concrete unit.

puddled clay – clay worked to a suitable consistency to form a barrier to the passage of water.

purlin – one of several horizontal structural members that support roof loads and transfer them to roof beams.

railway gauge: the distance between the inner faces of the rails

standard Irish gauge – the track gauge of 5ft 3in in general use in Ireland;

narrow gauge – any gauge less than standard Irish gauge, usually either 2ft or 3ft.

reinforced concrete – concrete containing steel reinforcement, thus increasing its load-bearing capacity.

rolling lift (Scherzer) – a type of bascule developed by William Scherzer of Chicago in 1893. As well as lifting, it also rolls back at the pivot point.

shear or **shear force** – a transverse force acting on a structure.

spalling – splitting off, particularly of concrete

spandrel – the triangular area between the arch and the deck of an arch bridge.

stiffener – a rib-like projection from a thin structural member which increases the stiffness and thereby prevents buckling due to compression.

string course – a projecting course of masonry or brickwork, often framing an arch or below the parapet of a bridge.

suspension – in suspension bridges the deck is supported by hangars from a suspension cable supported by towers and anchored to a foundation behind them..

swing bridge – a bridge which swings open horizontally to allow a ship or barge to pass.

tension – a pulling force.

tide mill – a mill operated by water from millpond filled by the incoming tide.

tie bar – a metal rod inserted to counter movement apart of two structural elements.

tower mill – a windmill with a brick or stone tower. The top of the tower containing the sails rotates, the machinery in the tower remains stationary.

truss – a girder in which the web is formed of discrete vertical and/or inclined members. There are many different configurations.

viaduct – a multi-span bridge used to carry a road or railway over a valley, or over other roads or railways.

voussoir – the individual shaped blocks forming the arch.

wallower – the main gear in a wind or water mill which converts the rotation in a vertical plane of the sail or wheel into rotation in a horizontal plane to drive the machinery.

water wheels:

breast shot wheel – supplied by water at about mid-height;

pitch back wheel – fed at the top but turning in the opposite direction to the flow, more efficient than an overshot wheel;

overshot wheel – supplied by water at the highest point of the wheel and turns in the direction of the flow;

undershot wheel – supplied with water at the bottom of the wheel.

weir – a dam in a stream or river to raise the water level or divert its flow.

wrought iron – an iron-carbon alloy with few impurities capable of being rolled or forged as plate or bar.

BIBLIOGRAPHY

Barry, M.B. (1985) *Across Deep Waters*. Dublin, Frankfort Press

Bielenberg, A. (ed.) (2002) *The Shannon Scheme and the electrification of the Irish Free State*. Dublin, The Lilliput Press

Broderick, D. (2003) *The First Toll-Roads: Ireland's Turnpike Roads 1729–1858*. Cork, The Collins Press

Casserley, H.C. (1974) *Outline of Irish Railway History*. Newton Abbot, David and Charles

Clarke, P. (1992) *The Royal Canal: The Complete Story*. Dublin, ELO Publications

Corcoran, M. (2005) *Our Good Health: A History of Dublin's Water and Drainage*. Dublin, Dublin City Council

Cox, R.C. & Gould, M.H. (1998) *Civil Engineering Heritage: Ireland*. London, Thomas Telford Publications

Cox, R.C. & Gould, M.H. (2003) *Ireland's Bridges*. Dublin, Wolfhound Press

Cox, R.C. (ed.) (2006) *Engineering Ireland*. Cork, The Collins Press

Delany, R. (1992) *Ireland's Inland Waterways*. Belfast, Appletree Press

Delany, R. (1992) *Ireland's Royal Canal*. Dublin, The Lilliput Press

Delany, R. (1995) *The Grand Canal of Ireland*. Dublin, The Lilliput Press

Delany, R. (2008) *The Shannon Navigation*. Dublin, The Lilliput Press

Donnelly, K., Hoctor, M. and Walsh, D. (1994) *A Rising Tide: The Story of Limerick Harbour*. Limerick, Limerick Harbour Commissioners

Fayle, H. (1946) *The Narrow Gauge Railways of Ireland*. London, Greenlake Publications

Feehan, J. and O'Donovan, G. (1996) *The Bogs of Ireland*. Dublin, Environmental Institute, University College Dublin

Ferris, T. (2008) *Irish Railways: A New History*. Dublin, Gill & Macmillan

Ferris, T. (2010) *The Trains Long Departed: Ireland's Lost Railways*. Dublin, Gill & Macmillan

Flanagan, P. (1994) *The Shannon–Erne Waterway*. Dublin, Wolfhound Press

Gilligan, H.A. (1988) *A History of the Port of Dublin*. Dublin, Gill & Macmillan

Hague, D.B. and Christie, R. (1975) *Lighthouses, Their Architecture, History And Archaeology*. Llandysul, Gomer Print

Ireland, John de Courcy (2001) *History of Dun Laoghaire Harbour*. Dublin, deBurca

Krauskopf, S. (2001) *Irish Lighthouses*. Belfast, Appletree Press

Lohan, R. (1994) *Guide to the Archives of the Office of Public Works*. Dublin, Stationery Office

Long, W. (1997) *Bright Light, White Water*. Dublin, New Island Books

Louden, J. (1940) *In Search of Water*. Belfast, W.H. Mullan & Sons

Manning, M. & McDowell, M. (1984) *Electricity Supply in Ireland: The History of the ESB*. Dublin, Gill & Macmillan

Marmion, A. (1855) *The Ancient and Modern History of the Maritime Ports of Ireland*. London, The Author

McCutcheon, W.A. (1965) *The Canals of the North of Ireland*. Newton Abbot, David and Charles

McCutcheon, W.A. (1980) *The Industrial Archaeology of Northern Ireland*. Belfast, HMSO

Motorway Archive Trust (2002) *The Northern Ireland Motorway Achievement*. Belfast, NI Roads Service

Mulligan, F (1983) *One Hundred and Fifty Years of Irish Railways*. Belfast, Appletree Press

Murray, K.A. (1944) *The Great Northern Railway (Ireland): Past, Present and Future*. Dublin, GNR(I)

Murray, K.A. and McNeill, D.B. (1976) *The Great Southern and Western Railway*. Dublin, Irish Railway Record Society

O'Connor, K. (1999) *Ironing the Land*. Dublin, Gill & Macmillan

O'Keeffe, P.J. (2001) *Principal Roads in Ireland, AD 123 to 1608*. Dublin. The National Roads Authority

O'Keeffe, P.J. (2003) *Principal Roads in Ireland, 1608 to 1898*. Dublin. The National Roads Authority

O'Keeffe, P.J. & Simington, T. (1991) *Irish Stone Bridges: History and Heritage*. Dublin, Irish Academic Press

Patterson, E.M. (1962) *The Great Northern Railway of Ireland*. Lingfield, Oakwood Press

Pearson, P. (1981) *Dun Laoghaire Kingstown*. Dublin, O'Brien Press

Robb, W. (1982) *A History of Northern Ireland Railways*. Belfast, Northern Ireland Railways

Rowledge, J.W.P (1995) *A Regional History of Railways, Volume 16: Ireland*. Penryn, Cornwall, Atlantic Transport Publishers

Ruddock, Ted (1979) *Arch Bridges and their Builders 1735–1835*. Cambridge, Cambridge University Press

Rynne, C. (2006) *Industrial Ireland 1750–1929*. Cork, The Collins Press

Scott, C.W. (1906) *History of the Fastnet Rock lighthouses*. Dublin, Commissioners of Irish Lights

Shepherd, W.E. (1974) *The Dublin and South Eastern Railway*. Newton Abbot, David and Charles

Shepherd, W.E. (1994) *The Midland Great Western Railway of Ireland*. Leicester, Midland Publishing

Shepherd, W.E. (2006) *Waterford, Limerick and Western Railway*. Hersham, Surrey, Ian Allan Publishing

Shepherd, W.E. (2005) *Cork, Bandon & South Coast Railway*. Leicester, Midland Publishing

Sweetnam, R. and Nimmons, C. (1985) *Port of Belfast 1785–1985*. Belfast, Belfast Harbour Commissioners

Taylor, R.M. (2004) *The Lighthouses of Ireland*. Cork, The Collins Press

Welbourn, N. (2006) *Lost Lines: Ireland*. Hersham, Surrey, Ian Allan Publishing

INDEX OF PERSONS

Arup, Ove & Partners 172, 229
Atkinson, William (*c.* 1773–1839) 185–6

Babtie, Shaw & Morton 254
Baker, Sir Benjamin (1840–1907) 61, 125
Bald, William (1789–1857) 11, 80, 201, 217, 221
Bazalgette, Sir Joseph William (1819–91) 78–9
Barrington, Sir Matthew (1788–1861) 199
Barrington, William (1825–95) 213
Barton, James (1826–1913) 55, 121
Bateman, John Frederick Latrobe (1810–80) 270
Bayley, Kennett (1838–1911) 193
Benson, Sir John (1812–74) 183, 191, 217
Bianconi, Carlo (Charles) (1786–1878) 10
Binnie, Deacon & Gourley 244
Bligh, William (1754–1817) 96, 148
Bond, Francis Willoughby (1901–1953) 164
Brand, Charles & Sons Ltd. 139, 244, 254, 257
Brennan, James M. (*c.* 1912–1967) 172
Bretland, Josiah Corbett (1846–1921) 258
Brindley, James (1716–72) 37
British Insulated Callender's Construction Co., Ltd. 236
Brownrigg, John (*c.* 1749–1838) 29–30, 33, 39, 115, 206, 240
Brunel, Isambard Kingdom (1806–59) 59, 122, 191
Burden, Alexander Mitchell (1864–1923) 142
Burnett, Thomas 254
Butler, William Deane (*c.* 1794–1857) 58

Campbell, W.J. & Sons 256, 260
Castle (Cassells), Richard (1690–1751) 28
Cementation Co., Ltd. 134, 231
Chance, Norman Albert (*c.* 1885–1963) 134
Chapman, Edward (*fl.* 1792–96) 154
Chapman, William (1749–1832) 33–4, 36, 96, 112, 153–4
Chatterton, George (1853–1910) 79
Clarke, Eugene O'Neill (1867–1943) 211
Coffey, Andrew (?–1833) 173
Cooper, J.D. 271
Courtney, Stephens & Co. 126, 135, 160, 167, 246
Cox, Lemuel (1736–1806) 211
Cubitt, Sir William (1785–1861) 119, 159

da Vinci, Leonardo (1452–1519) 26
Daglish, Robert (jnr) (1808–1883) 169
Dargan, William (1799–1867) 42, 46,49–50, 55–6, 59, 161, 192, 198, 223, 241, 254, 256
Doherty, William James (1834–1898) 166–7, 271
Donald, Robert Buchan (1879–1963) 263
Doran, Dr I.G. & Partners 258, 262
Dorman Long & Co. 228
Dorman, John William (b. 1850, fl. 1909) 197
Douglass, William (1831–1923) 104, 180
Dredge, James (1794–1865) 238–9
Ducart, Davis (fl. 1761–1781) 37–8, 183, 230
Dunn, William Edmund L'Estrange (1843–1925) 200

Edgeworth, Richard Lovell (1744–1817) 11
Evans, Richard (fl. 1778–1802) 30, 34, 39–40, 112–4, 116–7
Evans, William 121
Expanded Metal Co. 249

Millar, Archibald (*fl.* 1786–1830) 36, 114, 153
Mitchell, Alexander (1780–1868) 141
Moir, Sir Ernest William (1862–1933) 242
Monks, Daniel (*fl.* 1779–1810) 30
Monks, Thomas 38
Moore, Courtney Edward (1870–1951) 123
Motherwell Bridge & Engineering Co. 121, 262
Mott, Hay & Anderson 228
Mouchel, L.G. & Partners 135, 142, 201, 230, 232, 235
Mouchel, Louis Gustave (1852–1908) 21
Moynan, John Ouseley (1850–1933) 212
Mullins, Bernard (1772–1851) 34,40,78,115–6,118, 254
Mullins, Michael Bernard (1808–71) 78
Mulvaney, John Skipton (*c.* 1813–1870) 50, 58, 159
Mulvany, William Thomas (1806–85) 43, 207
Myers, Christopher (1717–1789) 29, 37

Nash, John (1752–1835) 189
Neville, John (*fl.* 1840–89) 132
Neville, Parke (1812–86) 79, 132–3, 174
Nimmo, Alexander (1783–1832) 11, 18–19, 80, 110, 136, 158, 178, 187–8, 214, 217
Nixon, Charles (1814–73) 194

Oates, James (1778–1821) 153
Ockenden, William (*fl.* 1756–61) 33
Omer, Thomas (*fl.* 1750–70) 29–31, 33, 36, 111–2, 128, 206–7, 260–1
O'Sullivan, John George (1868–1912) 132
Otway, James (1843–1906) 125, 197
Owen, Jacob (1778–1870) 136
Owen, Richard (1744–1830?) 31, 44, 251

Page, Sir Thomas Hyde (1746–1821) 155
Pain, James (*c.* 1779–1877) 186, 189
Palladio, Andrea (1508–1580) 18
Papworth, George (1781–1855) 169
Parsons, William, 3rd Earl of Rosse (1800–1867) 130
Pearce, Sir Edward Lovett (1699–1733) 28, 240
Perronet, Jean Rodolphe (1708–94) 17–18, 187
Pollock, R.D. 249
Poncelet, Jean-Victor (1788–1867) 85
Price, Alfred Dickinson (1857–1934) 195
Price, James (1831–95) 236
Prior, Thomas (1681–1751) 80
Purcell, Pierce Francis (1881–1968) 164, 201

Quin, Peter 271

Rainey, John (Construction) 222
Ransomes & Rapier 227
Redpath Dorman Long 229
Rendel, Palmer & Tritton 254
Rennie, John (Senior) (1761–1821) 29, 85, 96, 98, 116, 150, 155–6, 158, 225, 265
Rennie, Sir John (1794–1874) 98, 136, 158, 225, 265
Rhodes, Thomas (1789–1868) 33, 128–9, 206, 209, 211
Rhodes, William (*fl.* 1790) 153
Ridley, M. Noel (1860–1937) 201
Riordan, Michael (*c.* 1879–1935) 183
Ritchie, Francis 258
Ritchie, William 98, 254
Rogers, Thomas (*fl.* 1788–97) 103, 155
Ryan, John Henry (1846–1929) 213

St Leger, Noblett R. (*fl.* 1834–72) 217
Salmond, Thomas Ross (*c.* 1829, *fl.* 1899) 254
Savage, James (1779–1852) 165
Scott, Michael John (1906–1989) 172
Searanke, Samuel S. (*fl.* 1849–68) 132
Semple, George (*fl.* 1753–82) 18, 83, 109, 167
Shepherd, Major W. Edward (1878–1948) 226–7
Simington, Thomas Aloysius (1905–1994) 164
Skinner, Andrew (*fl.* 1772–85) 9
Smeaton, John (1742–92) 34, 37, 85, 112, 265
Smirke, Sir Robert (1780–1867) 161

Smith, David 271
Smith, George (1783?–1869?) 18, 109
Smith, John Chaloner (1827–95) 163
Stanier, Charles Edward (1879, *fl.* 1928) 212
Steers, Thomas (1670–1750) 28, 30, 240
Stevens, Alexander (1730–1796) 165
Stevenson, Robert (1772–1850) 242
Stoney, Bindon Blood (1828–1909) 95, 121, 166–7

Tarrant, Charles (1815–77) 34, 111, 185
Taylor, George (*fl.* 1772–85) 9
Taylor, George J. (*fl.* 1809) 155
Telford, Thomas (1757–1834) 11, 42, 49, 51, 95, 149, 150, 187
Thompson, J. & R. 21, 135, 142, 249
Townsend, Edward (*fl.* 1866–90) 213
Trail, John (*fl.* 1756–87) 34, 111
Traill, William Acheson (1844–1934) 224
Trussed Steel Concrete Co. 234
Turner, Richard (*c.* 1798–1881) 58, 160

Vignoles, Charles Blacker (1793–1875) 49–51, 58, 158

Wallace, William (1883–1969) 227
Wejchert, Andrzej (Andrew) (d. 2009) 176
Whitworth, Robert (1734–99) 31
Wilkinson, George (1814–90) 58–9
Wimpey Asphalt 222
Wimpey, George & Co. 254
Windsor, John (*fl.* 1816) 110, 168
Wise, Berkeley Deane (1853–1909) 58
Wood, Sancton (1815–86) 58, 161–2
Worthington, Robert (*fl.* 1887–88) 213

Young, Arthur (1741–1820) 8–9, 46

INDEX OF PLACES